FRAMINGHAM STATE COLLEGE

3 3014 00054 9883

D1272452

RELATIVITY
REEXAMINED

Relativity
Reexamined

Léon Brillouin

 Academic Press *1970* *New York and London*

Framingham State College
Framingham, Massachusetts

COPYRIGHT © 1970, BY ACADEMIC PRESS, INC.
ALL RIGHTS RESERVED
NO PART OF THIS BOOK MAY BE REPRODUCED IN ANY FORM,
BY PHOTOSTAT, MICROFILM, RETRIEVAL SYSTEM, OR ANY
OTHER MEANS, WITHOUT WRITTEN PERMISSION FROM
THE PUBLISHERS.

ACADEMIC PRESS, INC.
111 Fifth Avenue, New York, New York 10003

United Kingdom Edition published by
ACADEMIC PRESS, INC. (LONDON) LTD.
Berkeley Square House, London W1X 6BA

LIBRARY OF CONGRESS CATALOG CARD NUMBER: 74-107560

PRINTED IN THE UNITED STATES OF AMERICA

QC
6
B72

Dedicated to Marcelle

this critical essay in a never finished science.

CONTENTS

7. A GRAVISTATIC PROBLEM WITH SPHERICAL SYMMETRY

8. REMARKS AND SUGGESTIONS

INDEX

Preface

After my last book on scientific uncertainty was published by Academic Press (1964), I felt relieved and free to go back to a variety of problems or dreams I had been pushing aside and postponing. Some problems seemed clearly stated but not solved, while many ideas remained rather cloudy or uncertain and required a good deal of thinking over. I decided to start looking into relativity which I wanted to examine from a new angle and with an unconventional perspective; traveling along high roads is no fun, but wandering on forgotten tracks may lead to some wild summit from which you suddenly discover the whole landscape with an uncommon beauty. All along my scientific career I felt an attraction toward problems arising along the border of a theory, in this little known territory where it joins a domain reserved to another theory. How does the first theoretical description check with the painting drawn from the second theory: How can you use either wave theory or discrete particles and obtain similar results, both agreeing with experiments? How do you know where to use geometrical optics or physical light waves? Coming back to relativity, where and how does it rejoin classical mechanics? Every example may reveal some curious situation: sometimes (but not too often) one of the solutions may appear as a convergent series, where the first term corresponds to the first theory, but this is not a general rule. Very often one may discover semi-convergent series, that can be used only up to a certain term and come close enough to the solution of the other theory, in the boundary region.

To form any notion at all of the flux of gravitational energy, we must first localize the energy. In this respect it resembles the legendary hare in the cookery book. Whether the notion will turn out to be a useful one is a matter for subsequent discovery. For this also, there is a well-known gastronomical analogy.

Heaviside—1893

Introduction

The value of a scientific theory lies in its ability to predict. In " Scientific Uncertainty, and Information "* it was emphasized that a theory T yields correct results with a certain maximum error ε only within a certain domain of applicability D. If one attempts to apply the theory too widely, outside its proper domain D, one shouldn't be surprised to obtain results with large errors. The boundary region, between domain D_1 of theory T_1 and domain D_2 of an adjacent theory T_2, is always very interesting to explore, and such an exploration may lead to the discovery of a variety of unspecified implicit assumptions made by the theoretician in his own field.

Every theory contains a number of quantities that can be measured by experiments and a few expressions that cannot possibly be observed. The first represent the *observables*, and the second are the *unobservables*. The distinction is not always made and many authors claim some data to be observable, according to arbitrary definitions, which do not correspond to any physical experiment. This leads to inconsistencies and paradoxes that should be avoided at all cost. Here I would take the strictest point of view and assume (after Bridgman) that an observable can be selected only if it corresponds to carefully described experimental equipment and method of observation.

If this is done, theory T_1, within its domain D_1, describes relations among its observables O_1, but also adds to this stock all

* Brillouin, 1964; hereafter referred to as SU.

sorts of relations containing unobservables U_1. These additional relations may be useful for a " description " of the theory, but they have no scientific meaning. Along the boundary between domain D_1 and domain D_2, the relations between observables O_1 and O_2 should be in close correspondence, but there will be discordances between references to unobservables U_1 and U_2. These discontinuities, once recognized, do not matter much.

At this point we may raise a most important question: *How much confidence do scientific theories deserve?* The answer must be cautious enough: a good deal, but not too much! There are limitations to all our theories; they are good up to a certain limit and within certain boundaries. They do not represent " The truth, nothing but the truth ... ." Every theory is based on experiments that have been checked very carefully, but the result can only be stated " within possible errors " between fixed limits according to the best knowledge of the experimenter. There is always a possibility that a new, unpredictable cause of errors might be playing a role in a new experiment, or that the theory has been extrapolated too far from its domain.

Let me stop a minute for a short story: I was driving through New Mexico, some years ago, and found myself entering a charming little city, called " Truth or Consequences." I stopped in front of the sign at the city limits, and wondered, what sort of " truth " was this? Certainly not scientific truth, and I smelled the reeking smoke of old pyres, the stink of intolerance; I imagined how poor pagan natives had been mistreated or witches tormented.

Scientific truth should never be taken so seriously, and every scientist must be ready to accept some adjustment and correction of his pet theories. There is no absolute truth in science, and here I must state that I am thinking of experimental science. Mathematics is another story.

Some traditional sciences are a curious mixture of observations, coupled with interpretations based on the best theories, but with an extrapolation so far from actual experiments that one may feel shivering and wondering: How much wishful thinking, how much science fiction. It is splendid to discuss the creation of our world, but never forget that you are dreaming, and do not expect the reader to believe in any model, whether with a sudden atomic explosion or with a story expanding back and forth from $-\infty$ to

+∞. All this is too wonderful to be true, too incredible to be believable.

Excuse me for another story: As a student at the University of Paris I attended a splendid series of lectures by Poincaré. For many years all the lectures delivered by Poincaré were immediately transcribed by one of his students and published by Gauthier-Villars. The student in charge may have been Borel, Drach, Chazy, or some other clever young man who later made a name for himself.

In 1911, Poincaré did not select anybody for this special work. He was lecturing on cosmogony; he knew the theories to be too unreliable, and repeatedly emphasized that the theoretical explanations offered by various authors were definitely inadequate. We did not know where the heat radiated by the sun came from. We did not understand how stars were built and how they died. There was too much missing from our knowledge. Sometimes, Poincaré suddenly stopped talking and silently walked back and forth in front of the blackboard. Then he turned to the audience, brushed aside all his notes and started: " I just have a new idea. Let us try whether it works ... ." He would state his new point of view and start working on the blackboard, computing numerical values, and conclude: " This is not much better than other theories; there is undoubtedly too much missing." This was the last complete series of lectures by Poincaré, just a year before his untimely death.

Do you think we are in a better situation now? We have certainly learned a great deal during the last fifty years. But we still are very far from understanding cosmogony. It remains a dream, a wonderful and evading dream.

Here some reader may say: We must trust some well established principles of symmetry in space and time, the principle of relativity, etc. Let us sketch now the *relativity of the principle of relativity*! This famous principle was first discovered in classical mechanics: The *laws of motion*, stated for a frame of reference at rest, remain *exactly the same when the motions are observed from a frame of reference moving with a given constant velocity v.* The reader may take notice of the fact that I speak of " frames of reference " instead of " sets of coordinates." There is a fundamental distinction to be made between the definitions, as we shall see in Chapter 4. A set of coordinates is a purely geometric definition; the coordinates have no mass, for the simple reason that geometry completely ignores masses. A frame of reference must have a mass, and this mass must be assumed to be much greater than that of any object moving within the frame.

For the moment, let us concentrate on the word " given." What do we mean by a *given velocity*? Who is giving us this velocity, and how? I become very suspicious whenever I hear the word " given." There is only one occasion when it has a definite meaning; this is in the statement of a problem given by an examiner to some helpless students. In this situation the velocity is supposed to be *exactly* the given velocity, with no possible error or discussion. But in real life, this never happens. If I observe an unknown moving object in the sky, nobody can *give* me its velocity. Whether it be a star or a flying saucer, I have to *measure the velocity* by some experimental device. I may use optical signals, which will be reflected from the unknown object, to measure the delays, the Doppler shifts, etc. From these measurements, I can compute the velocity, but I should always be aware of the fact that these very experiments always perturb the motion. The velocity after observation is not the same as before observation. *Every experiment requires some coupling between the observer and the observed object,* and the object is not in the same state of motion after the observation has been made and the coupling removed. This is now a well-known fact, supported by many examples of quantum theory. In the measurement of a velocity, we use light signals containing so many photons. When reflected these photons push back the reflecting object (recoil effect) and change its velocity.

The *given velocity* is just a *myth of our imagination*. It is a traditional blunder, resulting from the illusion that " looking at something can do no harm." In the physics of the nineteenth century, such an assumption seemed obvious; it was taken for granted, without any discussion. Now we know better. The *frame of reference moving with a given constant velocity does not exist* and never did. What can be discussed is the problem of a *heavy* frame of reference, with such a large mass that the perturbation due to the measurement of the velocity is negligible; such a specification leads to many complications, as we shall see in Chapter 5. The old-fashioned principle of relativity is a dream; it represents only a limiting case, but may not, for instance, be used without much care when it comes to moving atoms, electrons, neutrons, photons, neutrinos, and all these new mysterious " particles " (we have no better word to qualify them) of very small masses.

Similar remarks apply to many principles recently put forward with most incomplete discussions of how the "symmetry," for instance, can actually be measured.

These are just a few examples to show how scientists' viewpoints have progressively changed, and how many new problems emerged, which even Einstein's genius was unable to foresee fifty or sixty years ago.

We witnessed the invention of atomic clocks of incredible accuracy, whose physical properties differ very much from the clocks Einstein imagined. This will be discussed in some detail in Chapter 3. Let us mention here a real difficulty resulting from internationally adopted definitions. The unit of length is based on the wavelength of a spectral line of krypton-86 under carefully specified conditions with accuracy 10^8 and the unit of time is based on the frequency of a spectral line of cesium with accuracy 10^{12}. Hence, *the same physical phenomenon, a spectral line, is used for two different definitions: length and time*, and the velocity c of light remains undefined and looks arbitrary. It should be stated, once and for all, whether a spectral line should be used to define a frequency or a wavelength, but not both!

The above definitions are supposed to be made on earth, where there is a certain gravity field; Einstein's relativity predicts some change in the units of length and time when measured in regions with different gravity fields. It also predicts a change in the velocity of light c. With the legal definitions of length and time it seems rather difficult to check experimentally such predictions. This raises a very real problem of metrology. The purpose of this monograph is to consider this as well as other questions arising since the formulation of relativity and quantum theories at the turn of the century. In Chapter 1 we will review the historical sequence of events which led to these theories. But let us consider first the development of scientific theories in general.

We presented in SU a general discussion about the meaning of scientific theories, using information methods, and we emphasized the personal role of the scientist; his task starts with the selection of an experiment that can be practically isolated from the outside world and described completely, thus allowing the possibility of repeating the experiment in other laboratories and checking the results of the first observations. The scientist also uses his imagination for building a theory that might connect together a certain number of experimental facts. He may, with the help of this theory, predict some new results that will or will not be checked by new experiments. If necessary, the theory will either be corrected or rebuilt in order to account for new empirical data.

Scientific knowledge is based on empirical fact and theoretical interpretation. Both grow together step by step in a remarkable symbiosis, which was discussed in Chapters III and IV of SU. The role of human imagination in the theories was carefully scrutinized by Lindsay (1967) in a brilliant paper published in *Physics Today*:

> Science is a game, in which we pretend that things are not wholly what they seem, in order that we may make sense out of them in terms of mental processes peculiar to us as human beings ... . Science is a method for the description, creation, and understanding of human experience.

REFERENCES

Brillouin, L. (1964). " Scientific Uncertainty, and Information." Academic Press, New York.

Lindsay, R. B. (1967). *Phys. Today* **20** (12), 23.

Chapter 1 **Quantum Theory and Relativity**

1. Quantum Theory

Both quantum theory and relativity originated at the turn of the century. Both are now considered basic to our present scientific thinking, but they offer completely opposite characters in the way they were built, and also in their historic development. A close comparison of these two theories is most interesting and we can learn much by scrutinizing their developments.

We have already sketched the development of quantum theory (SU, Chapter IV, p. 41). It came to life in the year 1900 when Max Planck published his first paper on a theory of blackbody radiation. This paper started with a classical discussion of electromagnetic waves, but suddenly introduced the idea of *quantized energy* needed for a statistical discussion: The formula for blackbody radiation (isothermal radiation) compared favorably with experimental results, provided the quantum of energy is taken to be $h\nu$, i.e., proportional to the radiation frequency ν, where h is Planck's constant. The first part of the paper relied on continuity (Maxwell's equation) but it ended with an irreducible discontinuity. Planck's first theory is summarized by the equation:

$$\Delta E = h\nu$$

$$E = nh\nu, \qquad n = 0, 1, 2, 3 \ldots \text{(integer)} \qquad (1.1)$$

Planck himself seems to have been very much disturbed by this strange duality in his theory. He attempted to rebuild it in a different way and obtained a different result (the second theory):

$$\Delta E = h\nu$$

$$E = (n + \tfrac{1}{2})h\nu, \qquad n + \tfrac{1}{2} = \tfrac{1}{2}, \tfrac{3}{2}, \tfrac{5}{2}, \tfrac{7}{2}, \ldots \text{(half-integer)} \qquad (1.2)$$

The change from integer to half-integer did not matter much for blackbody radiation at usual temperatures. The new feature, however, was of importance when temperature became very low, since it indicated the possible existence of a *zero-point energy* $\tfrac{1}{2}h\nu$. Observable quantities were similar in both theories, since zero-point energy is actually almost impossible to observe. Planck was still dissatisfied with the strange mixture of continuity and discontinuity, but very soon the proof was given (by Poincaré and Ehrenfest) that such a discontinuity of energy was absolutely needed for an interpretation of experimental data on blackbody radiation. There was no way out.

The finite value of Planck's constant h and its physical meaning were very much in discussion. At the first Solvay Congress in Brussels (1911), Sommerfeld noted that h had the dimensions of " action " in mechanics and suggested quantization of action in some problems. The idea was of great theoretical interest, but the examples selected by Sommerfeld were not very convincing; however, at the same congress, Langevin showed that Sommerfeld's quantum of action gave a quantization of magnetism, called *magneton*, when applied to an electron trajectory. Langevin found in 1911 the quantity we now call " Bohr's magneton." He was only off by a factor 2π, due to an unknown numerical coefficient in Sommerfeld's assumption. This might be called the Sommerfeld third quantum theory.

The fourth theory was presented in Bohr's first paper on the hydrogen atom (1913). I was a student in Munich that year and happened to be in Sommerfeld's office while he was opening an issue of *Philosophical Magazine*; he glanced at it and told me: " There is a most important paper here, by N. Bohr. It will mark a date in theoretical physics." Soon after this Sommerfeld started using his own " quantum of action " to rebuild a consistent theory of Bohr's atom.

This is how *quantized mechanics* (the fifth theory) was born, and why it progressed so fast. It was definitely Sommerfeld who discovered the importance of the $\int p\,dq$ action integrals, which paved the way for modern quantum theory.

We retraced the first five steps of quantum theory; but we cannot describe the explosive extension that followed: Experimental results came thundering down, and each time the theory had to be adapted or partly rebuilt. Spin, Pauli's exclusion principle, de Broglie waves and Schrödinger wave mechanics soon integrated with Born–Heisenberg's matrix computation, commutation rules, Dirac's electron, etc. All this can be read in the successive editions of Sommerfeld's book (1919). We might count dozens and dozens of successive changes, and each time we would find the same pattern, described in the introduction.

The familiar cycle: new experimental facts—theory rebuilt—observables being maintained but coupled with some new unobservables—new predictions, leading to new experiments, etc. illustrates the splendid symbiosis of thinking and experimenting which results in an endless escalation of knowledge.

This, in our opinion, represents the fundamental procedure of healthy scientific progress. Any halt in the development may mean a serious hidden obstacle that could necessitate a completely new theoretical scheme.

Certainly many more stages in the expansion of quantum theories will follow. Everything now is based on quanta. Problems that scientists of my age painfully discussed in our younger times are now being taught in the first year of university physics.

Many attempts at strict logical axiomatizations were proposed to please the theoreticians for a short time, only to be suddenly broken to pieces by some new discoveries. Dirac, von Neuman, and many others did their best to rule the flood and channel its waters, but the dams they erected were soon overrun. We are now waiting for some new bright ideas to solve the problem of hundreds of new " elementary particles," which may also be called " quantized waves," and this certainly will lead to many unexpected discoveries.

2. Relativity

The modern theory of relativity also got started with this century. When we look at this theory, we discover a very different picture not at all similar to quantum theory. First, we must draw a line of demarcation between " restricted relativity," as Einstein called it, and " general relativity." Restricted relativity is the strongest of the two branches; it rests upon a long history of physical and astronomical observations, which were summarized in the definition of an *inertial frame of reference*. The whole problem is very clearly stated by Sommerfeld (1952) at the beginning of his book on mechanics.

The laws of mechanics presuppose the existence of an *inertial frame*, an imaginary structure at rest, usually assumed to have its center on the sun with x, y, and z axes in the directions of known fixed stars. With respect to such a frame of reference, free moving particles move along straight lines with constant velocities when no external forces interfere. We immediately discover that any frame of reference moving with constant velocity with respect to the first one obtains similar properties, and we have a whole family of inertial frames. This is the *principle of relativity in classical mechanics*.

Physicists later discovered some unexpected results, and observed that this principle was also valid for all laws of physics. *No physical experiment can detect a uniform translation of the laboratory* (used as a frame of reference); but physical experiments performed in a laboratory may detect an accelerated motion, a rotation for instance (Foucault's pendulum, Sagnac's optical experiment, etc.). A great many experiments in electromagnetism, optics, and other fields proved the validity of this principle of restricted relativity.

Einstein introduced the adjective " restricted," because he later tried to extend the principle to more general situations, but this extension was recently criticized in different countries by independent scientists who found many weak points in Einstein's assumptions. Many things happened since Einstein worked out his theory at the beginning of the century. Quantum theories invaded all chapters of physics, including mechanics and optics. Some of Einstein's assumptions looked safe, but they are now open to discussion and must be reexamined very carefully. While quantum theory helped us to discover many new phenomena in physics, we still have very few experimental checks of general relativity; it is time to go back to the " brave old relativity " and revisit all its territory. Every physicist feels that the very few (altogether three) experimental checks are really a meager result for too much computation. General relativity is a splendid piece of mathematics built on quicksand and leading to more and more mathematics about cosmology (a typical science-fiction process). But let us return to restricted relativity.

In order to obtain the needed invariance of physical laws with respect to any linear change of frame of reference, it was necessary to modify the definition of such changes and to introduce a new transformation of space and time coordinates. This was done between 1895 and 1905 by Lorentz (see Sommerfeld, 1952, p. 13). The famous Lorentz transformation modifies x, y, z, and t in such a way that it keeps c, the velocity of light, unchanged. The law of addition of velocities makes c an impassable maximum. No velocity v can exceed c:

$$v \leqslant c \tag{1.3}$$

This results in the curious fact that energy and mass become synonymous:

$$E = mc^2 \tag{1.4}$$

Both relations (1.3) and (1.4) of course modify the usual laws of mechanics, but this is done in such a way that the laws of classical mechanics hold when velocities are very much smaller than c. This ensures a smooth junction of relativity with classical theory (SU, p. 44).

All preceding relations checked correctly with experiments and Eq. (1.4) became famous in connection with atomic transmutations and atomic bombs. The " first " or " restricted " relativity stands as a monumental discovery.

It left, however, many unanswered questions, the most serious one being about gravitation. Newton had assumed that gravitation propagates with an infinite velocity, an assumption which goes back to Galileo and even then appeared unreasonable to his contemporaries.

According to condition (1.3) we must admit that gravitation propagates with a velocity v_g smaller than c or at most equal to c

$$v_g \leqslant c \tag{1.5}$$

Einstein assumed

$$v_g = c \tag{1.6}$$

and it seems that the scientific community adopted this assumption as obvious.

The assumption is very far from obvious, however, since there is absolutely *no experimental measurement* of v_g; we shall discuss the problem in Chapter 3. We shall also discuss more completely the relativity of relativity theory sketched at the end of the introduction (see Chapter 4).

REFERENCES

Bohr, N. (1913). *Phil. Mag.* **26**, 476, 857.

Brillouin, L. (1964). " Scientific Uncertainty, and Information." Academic Press, New York.

Planck, M. (1900). *Verhandl. Deut. Physik. Ges.* **2**, 1937.

Planck, M. (1937). *Ann. Physik* **4**, 553.

" Solvay Congress," pp. 316, 403. Gauthier-Villars, Paris, 1911.

Sommerfeld, A. (1919). " Atombau und Spektrallinien." Viegeg, Braunschwieig.

Sommerfeld, A. (1952). " Mechanics." Academic Press, New York.
    At the end of p. 15 and on p. 16, Sommerfeld briefly summarizes the viewpoint of general relativity; but this is quite a different story.

Chapter 2   **Some Problems about Restricted Relativity**

1. Relativity and Potential Energy

Einstein's relation between mass and energy is universally known. Every scientist writes

$$E = Mc^2 \tag{2.1}$$

but the role of potential energy is not always clearly stated. We must investigate this situation carefully and try to understand what sort of difficulties are raised by such a revision. (Brillouin, 1964a,b; 1965.)

Let us consider a *physical body*, which we assume to be a closed structure, with an isolating boundary letting no energy trespass. It contains a certain energy E_0, that we may measure in a frame of reference where the body stays at rest. The internal energy may be chemical, mechanical, kinetic, or potential; it will change all the time from one type to another type; we state that this energy E_0 yields a rest mass M_0 according to Eq. (2.1).

When the physical body is in motion with a constant velocity $\mathbf{v}$, we obtain a new mass M, with an energy E, and a momentum $\mathbf{p}$:

$$E_0 = M_0 c^2, \qquad E = Mc^2, \qquad \mathbf{p} = M\mathbf{v}$$

$$M = \frac{M_0}{(1 - v^2/c^2)^{1/2}} \tag{2.2}$$

The change from M_0 to M accounts for the mass of kinetic energy.

The physical body may be moving in a static field of forces and obtain, at a certain instant of time, an external potential energy U. Everybody assumes the total energy to be represented by the formula

$$E_{\text{tot}} = Mc^2 + U \tag{2.3}$$

where U remains unchanged, despite the motion of the body at velocity v; this fact reveals that *one completely ignores any possibility of mass connected with the external potential energy.* If this external potential energy had any mass, this mass would somehow be set in motion by the displacement of the physical body, and this moving mass would obtain some kinetic energy. No provision for any such effect can be seen in Eq. (2.3).

We are thus in a strange situation, where the internal potential energy obtains a mass, while the external potential energy does not.

2. The Meaning of Potential Energy in Relativistic Theories

The definition of *potential energy* plays a prominent role in classical mechanics, but when we turn to relativity, this quantity is high on the list of concepts needing reappraisal. The original classical definition cannot be maintained, since it is based on " absolute time " and " infinite velocity of propagation " for signals. Many other definitions are in trouble for similar reasons: the third principle of Newton (equal *action and reaction* at any distance) and the notion of center of masses, etc.

How could we speak of equal action and reaction between the sun and the earth, for instance, when it takes about 8 min for a signal to propagate from one to the other? In 8 min the earth travels quite a distance, and the attraction of the sun is modified. If an explosion occurs on the sun, its action will be felt on the earth 8 min later, and the reaction on the sun will come back 16 min later! The problem of the reliability of potential energy definitions is actually a very acute one.

There are other difficulties raised by relativity in the definition of moment of momentum, or of moment of inertia, and more generally in the discussion of all *problems involving rotations*, that should be carefully reexamined.

Let us concentrate on problems of potential energy. There must be a way out of the trouble, because we know that *relativity joins smoothly with classical mechanics* when the following conditions are fulfilled:

    a. All velocities v must be very small compared to the velocity of light c:

$$v \ll c \tag{2.4}$$

    (This condition involves using small potential energies.)

    b. Distances r must remain small, so that delays in the propagation of signals may practically be considered as negligible:

$$\frac{r}{c} \ll \tau \tag{2.5}$$

where τ is a characteristic time interval for the motion under consideration, e.g., its period.

In the problem of sun and earth interaction, the first condition (a) is nearly fulfilled (except in Michelson's experiments), but the second condition (b) is not.

What must now be done is to investigate carefully a type of definition that can be used for a relativistic quantity which could replace potential energy, and reduce to potential energy in classical mechanics. We shall then be in a position to examine the space distribution of the new quantity and of the corresponding mass.

Before we discuss this problem we must consider another difficulty, resulting from traditional methods of classical mechanics. Many of these methods cannot be extended to relativity, and finally also had to be abandoned in quantum theories. Classical mechanics, with its *absolute time*, can state and discuss problems with any number of particles (say: $M_1, M_2, \dots, M_n$) located, at a certain instant t of absolute time, at $\mathbf{r}_1, \mathbf{r}_2, \dots, \mathbf{r}_n$. The potential energy is supposed to be any function $U(\mathbf{r}_1, \mathbf{r}_2, \dots, \mathbf{r}_n)$, and the problem is discussed in a mathematical space with $3n$ dimensions. Most theorems of classical mechanics are stated in this very general way.

Such a method is not applicable to relativistic problems, where each particle (coordinates x_n, y_n, z_n) obtains its individual time t_n in a given frame of reference; relativity is characterized by the use of a four-dimensional space-time.

The change in definitions is very serious and its consequences are many. For instance, let us consider a system of two particles interacting together: Shall we state that potential energy is located on the first particle? Should it be attributed to the second one? Or split between them? If *energy means mass*, where shall we locate the mass? This is a fundamental question which we have to discuss.

The question has been ignored or evaded frequently, because it does not always appear clearly in all problems. One of the two bodies interacting may be very much heavier than the other one, hence almost motionless, e.g., the earth attracting Newton's apple! Newton carefully stated his third principle: The apple, too, is attracting the earth! But many theoreticians forgot about it: The earth does not move (so they said), it creates a steady field of forces, and the apple is moving in this " given " field. As a result, these theories would assume no mass corresponding to potential energy, and write the total energy as in Eq. (2.2). The flaw is, however, obvious, and this is why the present discussion is needed.

3. The Importance of Fields in Einstein's Theories

All these questions hang closely together; they are tightly interrelated and have been considered by a great thinker like Einstein. He explained clearly that since action at a distance is forbidden, one should rely entirely on actions transmitted step by step by fields propagating through space. The importance of field theory was definitely brought into the foreground. The ideas launched by Faraday and Maxwell were completed by relativistic discussions. Fields were assumed to have a real physical existence, even when they do not act on any moving particle and go on unnoticed. Such an assumption looks pretty much like metaphysics, but it plays a dominant role in relativistic problems.

There is no longer any question of action and reaction at finite distances, but the law of equal action and reaction applies locally, at any given point *xyzt* in space-time.

The *field assumes a very complicated role*: It carries energy, momentum, Maxwell's tensions, etc., and we want to emphasize the fact that *the field itself carries a mass*. This is the situation which we intend to discuss, since its full significance has been partly overlooked by many theoreticians of relativity.

Let us start with a simple problem on which there is general agreement. We consider a sphere of radius a, with a mass M_0 and an electric charge Q that is distributed on the sphere's surface. In a frame of reference at rest, this charge Q generates an electric field $\mathbf{F}$ at a distance $\mathbf{r}$

$$\mathbf{F} = \frac{Q}{r^2}\mathbf{r}^0 \tag{2.6}$$

where $\mathbf{r}^0$ denotes a unit vector in the $\mathbf{r}$ direction. This electric field obtains an *energy density* (ESCGS units)

$$\mathscr{E}_{el} = \frac{1}{8\pi}\,|\,\mathbf{F}\,|^2 = \frac{Q^2}{8\pi r^4} \tag{2.7}$$

According to the fundamental rule (2.1), this corresponds to a *mass density*

$$\rho_m = \frac{1}{8\pi c^2}\,|\,\mathbf{F}\,|^2 = \frac{Q^2}{8\pi c^2 r^4} \tag{2.8}$$

The energy density (2.7) and mass density (2.8) can be integrated over the whole space, around the sphere a, and yield

$$E_{el} = \frac{Q^2}{2a}, \qquad M_{el} = \frac{Q^2}{2ac^2} \tag{2.9}$$

where E_{el} is the total electric energy in the field, and M_{el} represents the total mass in the field around the sphere. The sphere may have another mass M_0 of internal origin and its global mass amounts to

$$M_g = M_0 + M_{el} \tag{2.10}$$

When we write such a formula, we take into account the fact that Eq. (2.8) indicates a very high concentration of mass in the immediate neighborhood of the sphere, and we assume that this mass may (*as a first approximation*) be taken as located upon the sphere itself.

4. Two Interacting Spheres

Let us go on with electric problems that are better known than many other similar ones and can be used as typical examples. We now select a two-body problem, with two spheres of very small radius a, rest masses M_0 and M'_0, charges Q and Q', supposed *at rest in a certain frame of reference*; we call r_0 the distance between them. Let us call P a point in space (Fig. 2.1) where we observe the resulting electric field

$$\mathbf{F} = \frac{Q}{r^2}\mathbf{r}^0 + \frac{Q'}{r'^2}\mathbf{r}'^0 \tag{2.11}$$

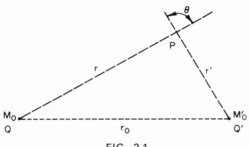

FIG. 2.1

The *electric energy density* is now given by the formula

$$\mathscr{E}_{el} = \frac{1}{8\pi}\,|\,\mathbf{F}\,|^2 = \frac{1}{8\pi}\left[\frac{Q^2}{r^4} + \frac{Q'^2}{r'^4} + 2\,\frac{QQ'}{r^2r'^2}\cos\theta\right] \tag{2.12}$$

where θ represents the angle between the vectors $\mathbf{r}$ and $\mathbf{r}'$.

The mass density becomes

$$\rho_m = \frac{\mathscr{E}_{el}}{c^2} = \frac{1}{8\pi c^2}\left[\frac{Q^2}{r^4} + \frac{Q'^2}{r'^4} + 2\,\frac{QQ'}{r^2r'^2}\cos\theta\right] \tag{2.13}$$

In this remarkable formula, the first term obviously represents the contribution to the mass M_0 of the first particle, while the second term contributes to the M'_0 mass of the second particle, but *what is the meaning of the third term, with the QQ' cross product?*

In order to clarify this point, let us first consider the integral of the cross product in formula (2.12) for electric energy. We call E_{int} the third term, that represents interaction between Q and Q'

$$E_{int} = \int_{-\infty}^{\infty}\mathscr{E}_{int}\,d\tau = \frac{1}{4\pi}\int(\mathbf{F}\cdot\mathbf{F}')\,d\tau$$

$$= -\frac{1}{4\pi}\int\left(\frac{\partial V'}{\partial x}F_x + \frac{\partial V'}{\partial y}F_y + \frac{\partial V'}{\partial z}F_z\right)d\tau$$

$$\tag{2.14}$$

where xyz are the coordinates of point P, while $d\tau$ is a volume element in space and $(\mathbf{F} \cdot \mathbf{F}')$ is the scalar product.

We introduce the static potential V' for charge Q', normalized by the usual condition $V' = 0$ at infinity,

$$V' = \frac{Q'}{r'} \qquad (2.15)$$

Integrating by parts, we find

$$E_{\text{int}} = -\frac{1}{4\pi} \mid V' (F_x + F_y + F_z) \mid_{-\infty}^{+\infty} + \frac{1}{4\pi} \int V' (\mathbf{\nabla} \cdot \mathbf{F}) d\tau.$$

The integrated term is zero and in the integral we note

$$(\mathbf{\nabla} \cdot \mathbf{F}) = 4\pi\rho_{\text{el}} \qquad (2.16)$$

where ρ_{el} is the electric density for charge Q. The result is

$$E_{\text{int}} = V'Q = \frac{Q'Q}{r_0} \qquad \text{assuming} \quad a \ll r_0 \qquad (2.17)$$

hence we have the following theorem:

The integrated interaction energy, taken over all space, yields the quantity usually called " potential energy " for two charges Q and Q', at rest in a certain frame of reference.

This means also that the total mass due to the cross-product energy terms QQ' represents the mass of potential energy and is actually distributed in the whole space

$$M_{\text{pot}} = \frac{Q'Q}{r_0 c^2} \qquad (2.18)$$

For two point charges QQ' *at rest* in a certain frame of reference, we have been able to replace the mathematical abstraction of potential energy by a physical model, where the energy is distributed in space according to the field.

If we now want to discuss a problem of moving charges, we have to follow a similar procedure and compute the energy density in the field of both interacting particles. Terms in QQ' will yield directly the interaction energy, for any distance and any velocity. The energy distributed in space, according to the field, corresponds to distributed mass.

Let us, for instance, consider a problem with one charge Q' at rest in a certain frame of reference, and the other mass moving with a velocity v. The field of Q' is the static field $\mathbf{F}'$ of Eq. (2.6), but the field $\mathbf{F}$ of the moving charge Q is represented by the well-known relativistic formulas [see, e.g., Sommerfeld, 1952, p. 240, Eq. (14)].

Terms in QQ' in the energy density may then be computed (in this special frame of reference) together with the corresponding mass distribution.

5. Where Could the Mass of Potential Energy Be Localized?

Let us consider a problem where conditions (2.4) and (2.5) are fulfilled and we can speak of potential energy.

The mass of potential energy is actually distributed in the whole space, between and around the charges Q and Q'. If, however, we look more closely into formula (2.13), we notice that the cross term (interaction)

$$\rho_{m,\,\text{int}} = \frac{QQ'}{4\pi c^2 r^2 r'^2} \cos \theta \qquad (2.19)$$

become very large on the charged spheres when either $r = a$ or $r' = a$. This indicates a concentration of mass near the two charges with much smaller density at a distance. The concentration, however, is not so strong as in Eq. (2.8); it goes as r^{-2} instead of r^{-4}. Nevertheless, we may introduce a first approximation similar to the one used in Section 2.3 and state: *For spheres of equal radii $a \ll r_0$, as a first approximation, the mass of potential energy can be considered as localized on the interacting charges QQ' and split 50/50 between them.* We rewrite Eq. (2.10) for the global masses in the following way:

$$M_g = M_0 + M_{\text{el}} + \frac{QQ'}{2r_0 c^2}$$

$$M'_g = M'_0 + M'_{\text{el}} + \frac{QQ'}{2r_0 c^2} \qquad (2.20)$$

The distribution of Eq. (2.19) is completely symmetrical in r and r' and this justifies the 50/50 split when boundaries are symmetrical too.

Some details, however, are worth discussing (Fig. 2.2). Formula (2.19) shows that the density of mass (and energy) obtains a certain sign at large distance, when θ is small and $\cos \theta$ is nearly unity. The $-$ or $+$ sign at large distance is given by the sign of

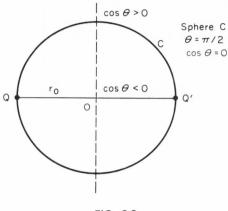

FIG. 2.2

the product QQ' and is the same as the sign in Eq. (2.18). However, we must notice that the $\rho_{m,\,int}$ density (2.19) is zero on a sphere C of diameter QQ', where we have $\theta = \pi/2$ and $\cos \theta = 0$. Within the sphere C, the density $\rho_{m,\,int}$ has an opposite sign.

Anyhow, the densities $\rho_{m,\,int}$ may have $+$ or $-$ signs, and (just as the potential energy itself) the mass of potential energy can be positive or negative.

The new masses (2.20), computed for particles at rest should be a good first approximation when one of the particles moves at a low velocity v, and corrections should be only in v^2/c^2.

6. Many Interacting Charges at Small Distances and Small Velocities

We discussed in some detail the case of two interacting electric charges Q and Q'; the results can be generalized to dipoles, quadripoles, or multipoles interacting with an electric charge.

Let us, for instance, consider a rigid structure at rest, holding a certain number of charges Q', Q'', ... , $Q^{(n)}$ and acting upon a free charge Q. This may, for example, correspond to the problem of a crystal lattice, with a free electron Q moving through the lattice. The charges Q', Q'', ... , $Q^{(n)}$ are supposed to have equal radii a; they have electric interactions among themselves, and this interaction will be part of the total potential energy (and mass) of their rigid structure. The free charge Q (also of radius a) interacts with any one of the $Q^{(n)}$ charges, and half of the corresponding mass of interaction is localized on Q, while the other half is on $Q^{(n)}$. Let us call U the potential energy of all these interactions:

$$U = \sum_{j=1}^{n} \frac{Q\,Q^{(j)}}{r_j}, \qquad a \ll r_j \qquad (2.21)$$

The mass of the free charge Q interacting with the structure becomes

$$M_Q = M_0 + M_{\text{el}} + U/2c^2 \qquad (2.22)$$

while there is an additional $U/2c^2$ mass on the rigid lattice. This is a straightforward generalization of Eq. (2.20).

Let us now assume the charge Q to be moving with a small velocity v; then the total energy of particle Q plus lattice is

$$E_{\text{tot}} = \frac{M_0 + M_{\text{el}} + U/2c^2}{(1 - v^2/c^2)^{1/2}} c^2 + \tfrac{1}{2}U \qquad (2.23)$$

instead of (2.3).

This can be rewritten in a slightly different way:

$$E_{\text{tot}} = \frac{M_0 + M_{\text{el}}}{(1 - v^2/c^2)^{1/2}} c^2 + U + \left[\frac{U}{2} \left(\frac{1}{(1 - v^2/c^2)^{1/2}} - 1 \right) \right] \qquad (2.24)$$

The last term within the brackets is the new term corresponding to our theory, as shown directly by a comparison of (2.24), (2.3), and (2.10).

In most practical applications this new term remains small, and Einstein's equation, (2.3), represents a good approximation. Our new correction might become of importance only for large values of the velocity v and of the potential energy U; but a large velocity v would require special treatment as noticed at the end of Section 2.4. According to the sign of U, the correction may be positive or negative. In such discussions, one should always beware of so-called potentials, that are usually defined up to an arbitrary constant (or function) and directly lead to " gauge " troubles.

The assumption that the new mass distribution is primarily located on the electric field in the whole space satisfies the obligation for relativistic transformations just as for the electromagnetic field itself. The simplified model with additional mass localized on the particle must be considered only as a simplifying approximation, enabling us to establish the junction with classical problems.

7. Unequal Particles; Role Played by the Geometry of the Boundary

In Sections 2.4–2.6 we assumed that all the interacting particles were spheres of radius a, and this led directly to condition (2.20) with the 50/50 splitting of the additional mass of two interacting particles.

Let us now discuss the more complex problem of two unequal particles. It is immediately obvious that the charges Q and Q' appear only by their product QQ'. The symmetry of the field of interaction remains unsensitive to any difference between Q and Q'. The masses M and M' do not seem to play any role either, but we shall come back to this point later. The field distribution is perfectly symmetrical with respect to the locations of the charges, but the *boundary conditions* depend on the *radii a and a' of the spheres*. We carefully specified that our first discussion required

$$a = a' \ll r_0 \qquad (2.25)$$

If a is different from a', the whole symmetry is broken.

At the same time, the masess M and M' will be different since their electric parts are different. For instance, let us assume $a > a'$. We obtain, according to (2.9)

$$a > a', \qquad M_{\mathrm{el}} < M_{\mathrm{el}}' \qquad (2.26)$$

Since the domain of integration is dissymmetrical, we cannot predict the 50/50 splitting of the additional mass of potential energy. The field is weaker around particle a of smaller mass M; also we will be inclined to give less interaction mass to this particle, and to replace Eq. (2.20) by

$$M_{\mathrm{int}} < \frac{QQ'}{2r_0c^2} < M_{\mathrm{int}}' \qquad (2.27)$$

Let us reexamine the discussion of Eq. (2.19) dealing with Fig. 2.2. The mass density ρ is just $\rho_{int}\, c^{-2}$ and takes opposite signs in different regions of space. This will create a field of gravitation corresponding to a gravity multipole, not to a single pole (single mass). These conditions may lead to a gravity potential of tensorial character, such as the one obtained by Einstein or by Dicke. If, instead of spherical particles, we consider charged particles of different shapes, the symmetry of the boundaries is completely destroyed and the 50/50 splitting of the interaction mass looks unreasonable.

Let us consider, for instance, the problem of one spherical particle located *inside* a closed metallic box. This box may be connected to a van de Graaf generator and maintained at a high potential V. This raises a very real problem because the mass of the interaction energy may be many times greater than the mass of the particle. For an electron, we have

$$m_0 c^2 \approx 500{,}000 \text{ eV}$$

However, we may choose

$$V = 10 \text{ MV}, \qquad U_0 \approx 20\, m_0 c^2$$

This is no longer a small correction; but here we have a complete dissymmetry. Let us start from the beginning when the box is empty; there is no charge inside, but a large charge Q and a large electric field is spread around the outside of the box. The energy of this field will simply add a contribution to the initial mass of the box.

Now we introduce one electron in the box. The field extending from the electron to the internal surface of the box is alone; there is no cross product in the electrical energy of the field, hence no cross energy and no interaction term. The field around the electron is the same as for a free electron in vacuum and the electrical mass of the electron is not perturbed. A surface charge density appears on the inner surface of the box, with a total charge exactly equal and opposite to that of the electron. A similar charge density (equal to one electron charge) appears on the outer surface of the box. The outside field is increased; its energy increases; its mass increases. Here there is no doubt that the whole mass of interaction is located on the box and *practically* no mass change can reach the electron inside. Complete dissymmetry achieves a situation where all the additional mass of potential energy is on the electrodes and apparatus with no contribution to the mass of the electron *as shown by experiments*. Under such conditions Eq. (2.3) is valid, but this result is not obvious and it may not be a general result. This proves also that the assumption of localized masses is a very crude approximation. A private discussion with Dicke was very helpful in clarifying this situation.

8. Generalizations; Quantum Problems

We must be cautious about the difficulty of defining a potential energy and take care not to confuse it with so-called potentials (electric or vector) commonly used in electromagnetism. These potentials represent a four-vector, and may depend on x, y, z, and t; they are defined up to an arbitrary function and they have no direct physical meaning. Only their derivatives have a physical meaning and constitute the field components. It would be meaningless to connect the total energy of a system with the four-vector potentials. It is well known that these vector potentials lead to problems of " gauge invariance " and many complicated troubles.

We assumed in the preceding discussion that we had to deal with a *static problem* (in some preferred frame of reference), where the potential energy at infinite distance could be taken as the zero of potential energy, thus eliminating even an arbitrary constant in its definition. Our potential energy was a function of x, y, and z, but not of time t, defined in the preferred frame of reference.

Quantum problems were discussed by Lamb, Bethe, Schwinger, and others, and their papers can be found in Schwinger's book entitled " Quantum Electrodynamics " (1958). The method leads to corrections on the test mass of particles, called " mass renormalization," and yields excellent numerical results. Quantum effects include electrostatic potential energy and all sorts of spin effects.

The present discussion proves that *mass renormalization* is not only needed in quantum theories, but that it must already be introduced in classical relativity, where it was completely overlooked by the founders of relativity. Sommerfeld and Dirac were not aware of the difficulty, and their formulas must be very carefully revised.

9. Problems Arising at the Junction of Classical and Relativistic Mechanics

The problem we discussed was a typical example of the difficulties arising at the junction of two different theoretical models. Such problems were examined in a general way in SU, Chapters III, IV, and V, and the one we are discussing here is of special interest, since some of its peculiar characters seem to have been overlooked by the founders of relativity.

The junction between relativity and classical mechanics can be considered from two different viewpoints:

a. It was generally taken for granted that *relativistic mechanics* should reduce to *classical mechanics* when the velocity of light c could be *made infinite*. This may be mathematically correct for relativistic mechanics of particles, but this kind of reasoning is *physically unsound*. We may make c infinite in mechanics but we cannot assume anything similar in electromagnetism. The physicist, whether he is an experimenter or a theoretician, *cannot modify the velocity of light*. This velocity c is a fundamental constant in physics. When we speak of " mechanics " in this section, it should be well specified that we are thinking only of " systems of particles." We include problems of atoms and molecules but no continuous medium with wave propagation.

b. What a physicist can do is to investigate the properties of mechanical systems of particles when dimensions and velocities remain small [Eqs. (2.4) and (2.5)]. In such systems the delays for the propagation of signals may be so small that they become negligible, even with the finite light velocity c.

Conditions (a) and (b) actually lead to very different consequences. Let us, for instance, consider the mass–energy relation (E given)

$$E = Mc^2, \qquad M = E/c^2 \tag{2.28}$$

In problem (a) the mass M goes to the limit zero when c is infinite. In problem (b), the mass M remains finite.

Conditions (a) may satisfy a mathematician if he is interested only in mechanics of particles, but a physicist cannot accept them under any circumstances.

Conditions (b) have a real physical meaning, and exhibit another serious advantage. They are consistent with low frequencies ν, hence very small quanta $h\nu$; when the energy E of the system is large in comparison to $h\nu$, we really obtain classical mechanics, where neither quanta nor relativity can play any serious role.

The definition of *potential energy* in classical mechanics is based on the assumption that delays remain negligible for the propagation of any signals. Such an assumption is consistent with either condition (a) or (b). In relativistic mechanics, delays may become large, and the original definition is inapplicable. This difficulty was overcome [Eqs. (2.4) and (2.5)] when it was proved that this type of energy should no more be considered as " potential," but became very much real, and could easily be recognized in the field of interacting particles. The proof was given for electric fields, but it obviously can be extended to most other fields.

The duality encountered in conditions (a) and (b) of this section is much deeper than appears at first sight. We actually have to deal with *two different brands of special relativism*:

a. *Special relativity applied to systems of particles* (where the *mass–energy* relation (2.1) is *used only for kinetic energy*, while potential energy obtains no mass at all): In the applications of this (a) theory, most authors use " given scalar and vector potentials V and A " without specifying how these potentials have come into being. We have discussed these problems in Section 8.

b. *Special relativity in electromagnetism*: Here we have a much more comprehensive treatment, very carefully specified by Einstein and others. The mass–energy relation (2.1) applies for any kind of energy and all equations are consistent with a finite value of c. The so-called potential energy of mechanics can be discovered in the energy of the electric field distributed in all space, around electric charges, and this is enough to prove that it must be given a mass, but it does not tell us where to localize this mass.

Another question is unavoidable: In classical mechanics, the mass is always positive. Energy, on the contrary, as soon as we have defined the zero of energy, can be either positive or negative. In classical mechanics, the choice of the zero of energy is of no great importance; but in relativity, the absolute value of the energy does play a direct role. It is absolutely needed in our energy–mass relation (2.1). We must admit the possibility of *negative masses*, which correspond to negative energies.

ACKNOWLEDGMENT

The text and figures of this chapter are based on those appearing in L. Brillouin, *Proceedings of the National Academy of Sciences*, Volume 53, Number 3, March 1965. Reprinted by permission.

REFERENCES

Brillouin, L. (1964a). *Compt. Rend.* **259**, 2361.
Brillouin, L. (1964b). *J. Phys. Radium* **25**, 883.
Brillouin, L. (1965). *Proc. Natl. Acad. Sci. U.S.* **53**, 475, 1280.
Schwinger, J. (1958). " Quantum Electrodynamics." Dover, New York.
Sommerfeld, A. (1952). " Electrodynamics." Academic Press, New York.

Chapter 3   Gravitation and Relativity Quantized Atomic Clocks

1. How Does Gravitation Propagate?

We raised the question of how gravitation propagates at the end of Chapter 1, and we noticed that only one thing was certain: Gravitation must propagate with a velocity never exceeding c:

$$v_g \leqslant c \tag{3.1}$$

while Einstein simply assumed, without any experimental proof, that

$$v_g = c \tag{3.2}$$

This requires an explanation.

Einstein wanted to reduce all physics to pure geometry; he thought that a conveniently curved space–time universe would provide an explanation for all physical laws from electromagnetism to gravitation. This was his avowed aim and he worked toward this goal for half of his life. In order to achieve this goal, he could not introduce two different velocities v_g *and* c in his theory.

But the goal was never reached. Einstein managed to interconnect curved geometry and gravitation in a brilliant way, but his *unitary theory*, as he called it, was never achieved. Many attempts did not succeed, either because of a lack of generality or an excess of generality that left too many unknown arbitrary conditions; hence it was impossible to unite this geometric theory with electromagnetism.

Very few physicists now believe in the possibility of building such a unified theory. If this prejudicial assumption is rejected, there is no reason to maintain condition (3.2) and we are left with the inequality (3.1) until experiments yield the much awaited answer.

Has experiment given any such answer? We are very much disturbed to say bluntly: No!

What is worse, we have learned that empty space is propagating all sorts of other waves, even in " perfect " vacuum: de Broglie and Schrödinger waves are moving around with all sorts of velocities. We do not have just one velocity c but an almost infinite number of possible velocities. How can we guess which one of these velocities might correspond to the propagation of gravitation? We may imagine:

a. Gravity propagated by *actual waves* with velocity $v_g \leqslant c$, provided (Laplace, Le Verrier) this velocity is high enough not to disturb the motions in the solar system. These motions were computed for v_g infinite, and too small a value of v_g might significantly modify the interaction between planets.

b. Instead of actual waves, we may have gravitation spreading around according to a *diffusion equation*. Equations for heat propagation or diffusion contain a term in $\partial/\partial t$ instead of the $\partial^2/\partial t^2$ of actual waves. It is known that such equations progress initially with very high (even infinite) velocity; this anomaly, of course, should be corrected in order to preserve relativity conditions, but otherwise it seems difficult to rule out a possibility of diffusion equation for gravitation.

c. We may even think of *de Broglie or Schrödinger waves*! It is hard to see why the ψ waves could not be responsible for the propagation of gravity. Each particle has its ψ wave, and its mass makes it an emitter of gravity waves: Why not assume that the ψ waves propagate gravity? It may look strange, but since we know absolutely nothing of gravity waves, it seems difficult to rule out such an assumption.

d. We may also assume, instead of waves, an emission of " gravitons " with unknown velocities v_g!

Einstein may be right, and I am personally inclined to think he made the correct choice, but we have no experimental proof. Half a century has elapsed since Einstein formulated his assumption—Fifty years, during which a great many experimenters worked hard at the problem and were unable to design any experimental measurement of this velocity. The situation is really very disturbing.

2. Gravitation and General Relativity

When it comes to " general relativity," many new difficulties should not be overlooked, and the experimental meaning of the theory is far from clear. Bridgman (1955) wrote: " Einstein did not carry over into his general relativity theory the lessons and insights which he himself had taught us in his special theory."

Operational analysis was first applied by Einstein (1905) in his famous discussions about the meaning of length and time measurements for two frames of reference in uniform motion relative to one another. This was the basis of special relativity. But when he attacked general relativity, Einstein did not follow a similar procedure; he attempted to guess how to introduce gravity laws in relativity and to obtain a finite velocity of propagation for gravity forces. Some examples of operational discussion were first used to suggest the *equivalence* of gravitation and acceleration fields, but later Einstein introduced a very heavy mathematical structure that goes much beyond any physical need. And the experimental proofs of the theory are very few. Similar remarks were presented by Dicke (1967).

The role played by arbitrary " frames of reference," by " rigid yardsticks," or " exactly similar clocks " is extremely confusing. Here again, let us note our agreement with Bridgman (1955, pp. 319 *passim*) whose discussion can be used as a program for future research. A painful and complete reappraisal is absolutely needed.

3. Atomic Clocks that Einstein Could Not Foresee

Let us start with an obvious weak point: the lack of definition of *ideal clocks*. Such a definition was impossible at the beginning of the century, before quantum theory and Bohr's atom were discovered. We now have a definition, based on Bohr's (1913) second condition:

$$\Delta E = h\nu \qquad (3.3)$$

relating the frequency ν (measured in a frame of reference where the atom is at rest) to the energy transition E in the atom. Next to this relation we rewrite the mass–energy relation:

$$\Delta E = \Delta(mc^2) \qquad (3.4)$$

Energy, mass, and frequency are just one physical entity. A perfect clock is assumed to be stabilized on the frequency ν, for which we shall select, according to international agreements, the most stable atom structure we know from experiment, an atom of cesium. The special spectral line to be used and the conditions of observation have been very carefully specified. This clock essentially represents a *frequency standard*. One may use frequency demultiplication techniques to produce subharmonics of lower frequencies. This is done by amplifying devices using lasers, in connection with nonlinear structures. The multiplication or demultiplication of frequency by electronic devices plays a role similar to the cogwheels of old-fashioned clocks. These technical structures were first invented for low frequencies and had been used originally for comparing low frequencies of mechanical devices with high frequencies of vibrating piezoelectric crystals. It was then discovered that a low frequency vibrator could be " locked in " with a quartz oscillator vibrating on the frequency of a very high harmonic of the mechanical vibrator. From the quartz vibrator, the method was progressively extended to higher frequencies, first in the low infrared, then to optical frequencies and finally to the beginning of the ultraviolet region.

Officially, it is stated that the reliability of the cesium clock reaches 10^{11}, that is an error of one second in thirty centuries, but it seems probable that the accuracy might be improved up to nearly 10^{13} (a millisecond per century).

A much higher accuracy is obtained with the Mössbauer effect, an emission of γ-rays from an atom maintained at rest in a solid crystal. This was the frequency standard used by Pound in his wonderful experiments at Harvard (1959–1965), with Snider and other co-workers (see Pound and Snider, 1965). The atom kept at rest in a heavy crystal may emit γ-rays without any recoil effect and the accuracy of these rays is 10^{16}, at least. This, however, cannot be connected at present with atomic clocks at atomic frequencies because we do not know yet how to build lasers and frequency changers from ultraviolet up to γ-rays. This is still a formidable obstacle but it may be hoped that technical improvements will enable us to make the connection in the near future and to build *Mössbauer clocks with a precision of* 10^{16} (a microsecond error per century).

Why did Pound need a frequency standard of such incredible accuracy? He wanted to check one of the predictions of Einstein's general relativity, the so-called gravity red shift. This effect was verified with an accuracy of 1 % for the very small variation of gravity from the bottom to the top of a tower only 22 meters high, and this success was hailed as a wonderful check of Einstein's theory. We shall discuss the matter in a later section and show that another explanation can be put forward. The prediction was perfectly correct, but it can be interpreted differently. The Mössbauer effect is also an excellent example of a requirement that is usually ignored: A frame of reference must be very heavy in order to remain at rest during a physical experiment. This important question is discussed in Chapter 4.

In the general discussions of Einstein and Minkowski variables of time which are arbitrarily defined under a great variety of conditions are considered. Let us specify the meaning of an atomic clock: It measures the *proper time* in a laboratory where the *atomic clock is at rest*.

4. Atomic Clocks Are Not Einstein's Clocks

The great importance of the definition of atomic clocks is due not only to their fantastic accuracy (the highest ever recorded in physics), but also to the fact that these clocks build a bridge between relativity and quantum theory. The definition gives a *physical basis* for any discussion of the behavior of clocks under all sorts of perturbations. Einstein attempted to guess how clocks might depend on gravity. We shall be able to discuss the problem from general rules of quantum theory.

We remark first that the atomic clock yields a very precise definition of *one certain frequency*. It represents a *frequency standard*. Einstein's clocks were supposed to emit extremely short signals and to measure accurately time intervals between signals emitted and received. In a word, *an Einstein clock was a radar system*, and its requirements were thus very different from those of a frequency standard. It is well known that in order to emit a very short pulse it is necessary to use a very wide band of frequencies, not just one frequency. Requirements for the precise emission of an actual pulse are much more stringent and much more complicated than those for maintaining a strict frequency standard.

Let us now reexamine our quantum and relativity conditions (3.3) and (3.4). They represent the *fundamental basis of all physical sciences*. We will not pretend to explain these relations. They are beyond our comprehension. No theory (at least at present) is able to tell us why and how such relations may be understood. These identities

$$\text{energy} \equiv \text{mass} \equiv \text{frequency}$$

$$E \overset{hv}{=} Mc^2 \tag{3.5}$$

with their two numerical constants c and h, are the summary of all the laws of physics and cannot be derived from any present theory or model. It is the point of departure, not the result of our thinking. The mystery of this trinity is still complete.

Bohr made two fundamental assumptions in his famous paper on the hydrogen atom (1913):

1. He stated some conditions defining stable energy levels.
2. He stated condition (3.3) for the frequency v emitted or absorbed at a transition from one energy level to another.

This second Bohr condition survived unchanged through all the turmoil of fifty-five years of fantastic scientific discoveries. It is no use to summarize again this incredible period in the history of science, but one may recommend to the reader a most remarkable paper by Weisskopf (1968). Let us emphasize that Bohr's condition (1) for stable energy levels has been modified hundreds of times since its invention. It still has to be readapted almost every year to new experimental discoveries, but all the fundamental laws obtained up to now agree on the following rules:

1. There are stable energy levels at all stages of physics although stability criterion may change, and the stability itself may be of unknown duration.
2. Bohr's condition (3.3) always gives the frequency of emitted or absorbed radiation.

Condition (1) is so important that we must discuss it right away and quote parts of Weisskopf's essay. This author reminds us of the existence of three stages in spectroscopy:

   i. Atomic and molecular spectroscopy, with frequencies up to X-rays (also called electron spectroscopy)
  ii. Nuclear spectroscopy including γ-rays and radioactivity
 iii. Energies of excited particles discovered with powerful accelerators or cosmic rays

All three stages yield systems of stable energy levels. Transition from one energy level to another may correspond to the emission of one particle of total energy ΔE (rest mass M_0 plus kinetic energy) or to an emission of photons or neutrinos with zero rest mass.

Let us illustrate these statements with some figures borrowed from Weisskopf's brilliant paper. Figure 3.1 represents the energy

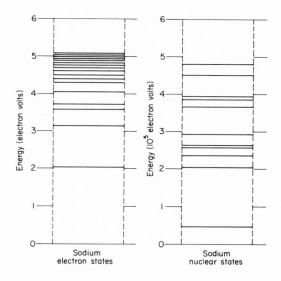

FIG. 3.1 Atomic and nuclear spectra of sodium are similar in character. But the atomic spectrum (*left*) can be plotted on a scale whose units are electron volts, whereas the spectrum of nuclear states (*right*) requires a scale whose units are larger by a factor of 100,000. Based on a figure from V. F. Weisskopf, " The Three Spectroscopies." Copyright © May 1968 by Scientific American, Inc. All rights reserved.

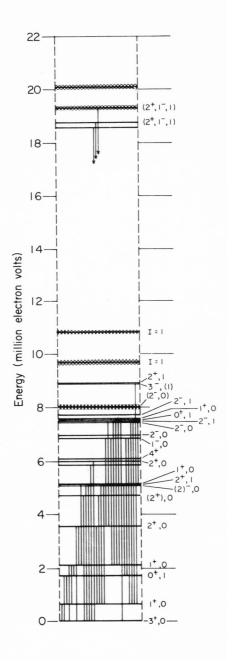

levels of sodium for electron states (on the left) and for nuclear states (on the right). What is striking, and was actually unexpected, is the fact that in both cases we discover sharp energy levels, and that transitions from one energy level to another yield a characteristic frequency of the atom. Electron states were calculated with quantum theory of outer electrons, while the nuclear states result from quantization of proton and neutron masses within the nucleus; the latter instance is a much more difficult theoretical problem. Nevertheless, we obtain sets of discrete energy levels of similar characters, and this is the point to emphasize in connection with our discussion on atomic clocks.

The extraordinary wealth of information included in such diagrams can be seen in Fig. 3.2, where nuclear energy levels for boron-10 are represented and a number of transition lines have been drawn (more than thirty of them), corresponding to high-energy photons emitted.

In addition to these diagrams referring to atomic spectroscopes (electron states) or nuclear spectroscopy, we have in Fig. 3.3 a very typical diagram of excited states obtained for high energy particles created in powerful accelerators. Here again, we must admire the appearance of well-defined energy levels and the great variety of transitions experimentally observed. This is a remarkable example of quantum classification, but for the moment we do not have any complete theoretical scheme to explain these extraordinary levels. The reader is referred to Weisskopf's paper for more information.

Let us conclude: *Discrete and well-defined energy levels* are the universal rule in atomic, subatomic, and even fundamental particle levels. The explanation of these energy levels and their theoretical interpretation are not yet completed.

FIG. 3.2 Nuclear Spectrum of boron-10 shows the principal transitions (*vertical lines*) in which high-energy photons are emitted. The first digit at the right of each quantum state is the spin angular momentum, the next symbol (+ or −) is the parity, the second digit is the isotopic spin *I*. Values in parentheses are uncertain. Gray bands indicate levels that are particularly broad. The figure follows one published by Thomas Lauritsen of the California Institute of Technology and Fay Ajzenberg-Selove of Haverford College. Based on a figure from V. F. Weisskopf, " The Three Spectroscopies." Copyright © May 1968 by Scientific American, Inc. All rights reserved.

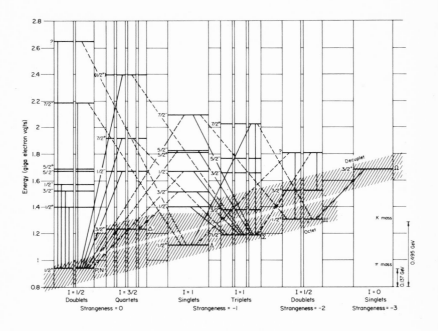

FIG. 3.3 Baryon Spectrum is composed of the nucleon (*P, N*) and its various excited states. The states are arranged in columns according to their multiplicity and strangeness. The letter *I* denotes isotopic spin; the multiplicity is given by $2I + 1$. Strangeness is an intrinsic quantum property. In the subnuclear spectrum of the baryon the ground state is taken to be the mass energy of the proton, 0·938 GeV. The number to the left of each state indicates spin angular momentum and parity (+ or −). The symbol to the right is the name of the state. The quanta emitted in certain transitions are shown in the key. Photon emissions are omitted; they generally link the same states linked by pions if there is no change in charge. Dashed lines indicate transitions that are mediated by weak interactions: lepton pairs or weak pion emissions. Transitions go from every member of a multiplet to every member of another, but for simplicity only one such transition is shown for each pair of states. The masses of pions and kaons appear at the right. The states in the octet and decuplet exhibit certain internal symmetries. Each baryon state shown here also exists in an antimatter state, so that there is a similar spectrum of antibaryons. Based on a figure from V. F. Weisskopf, " The Three Spectroscopies.'' Copyright © May 1968 by Scientific American, Inc. All rights reserved.

5. Accuracy and Reliability of Quantized Atomic Clocks

Atomic clocks based on Mössbauer's γ-rays have not yet been built, but they will eventually come into existence, because they offer the possibility of highest possible accuracy.

Atomic clocks using the best optical spectral lines are able to yield an accuracy keeping errors below 10^{-12} or 10^{-13}. This means errors smaller than a millisecond per century.

Mössbauer's γ-rays, as used by Pound, reach much further; errors are less than 10^{-16} meters or a split microsecond per century. In order to build a clock using these extremely fine lines, a great many technical difficulties would have to be overcome: First, a whole set of frequency multipliers or demultipliers covering the

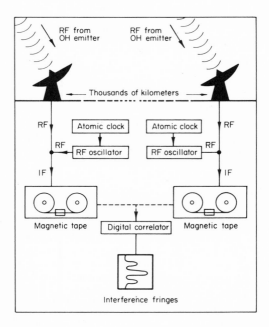

FIG. 3.4 Long-baseline interferometer uses two radio telescopes thousands of kilometers apart. Magnetic tape recordings synchronized by atomic clocks are correlated by computer to show interference fringes. [Reprinted from R. H. Dicke, *Physics Today*, November 1967, **20,** page 69.]

range from optical frequencies to γ-rays would have to be developed. This would mean a complete set of lasers and of nonlinear optical devices which do not extend beyond the ultraviolet at present. Let us hope that this extension will be realized without too much delay. It would enable us to perform many important experiments that would tell us definitely what to think of relativity!

The present clocks based on optical frequencies already display a remarkable accuracy. Let us consider, for instance, some problems of radio astronomy, and look at Fig. 3.4. Hydroxyl radicals (OH) from distant stars are found to emit radiation near 1665 MHz under strange conditions that puzzle astronomers. These rays are emitted from some areas near very hot stars, called HII regions, where the hydrogen is almost completely ionized. These regions are close to the galactic equator. There are four OH lines, observed in absorption and emission, and arising from hyperfine splitting of rotational levels. A curious feature is that the relative intensities of the four components are not in agreement with quantum theory. From Doppler effects, one observes that the OH groups are moving toward the center of the galaxy at about 40 km/sec. In contrast, H atoms are moving away at velocities of 50 km/sec. The physical size of the emitters is so small that it requires very long baseline interferometers, using two radio telescopes at a distance of many thousands of kilometers, from California to Norway. It is impossible to interconnect directly those stations, but they may be controlled and synchronized by two atomic clocks, and records on magnetic tapes can be compared by computer to show interference fringes. This is an extraordinary achievement and proves the splendid reliability of atomic clocks.

REFERENCES

Bohr, N. (1913). *Phil. Mag.* **26**, 476, 857.
Bridgman, P. W. (1955). " Reflections of a Scientist." Philosophical Library, New York.
Dicke, R. H. (1967). *Phys. Today* **20**, 55–70.
    Many references are listed in this article.
Einstein, A. (1905). *Ann. Physik.* [4] **17**, 133–138.
Pound, R. V. and Snider, R. L. (1965). *Phys. Rev.* **140**, B.788.
Weisskopf, V. F. (1968). *Sci. Am.* **218** (5), 15.

Chapter 4　**A Badly Needed Distinction between Mathematical Sets of Coordinates and Physical Frames of Reference**

1. Introduction: The Opinions of Bohr and Poincaré

Experimental science started with classical mechanics. We still draw from it major parts of our ways of thinking and of our definitions, and we use its language to express the results of our experiments, since these experiments are all conducted with laboratory instruments built on a human scale.

Let us quote here a very clear stand taken by Bohr (1958a) on the subject.

> The main point is to recognize that the description of experimental instruments, and the results of the observations, have to be expressed in the usual language of physical terminology along with its usual refinements. This is a simple, logical necessity, for the word " experiment " means only one process of which we can communicate to others what we have done and what we have learned.

Experimental equipment is built of strong and rigid material, heavy enough so that its position and speed can be determined in an absolutely classical manner without any possible intervention of the uncertainty principle of Bohr and Heisenberg.

In other papers Bohr (1958b) very clearly explains this point of view, which leads him to a presentation of his ideas of " correspondence " and " complementarity."

Without considering again these classical discussions, we would like to explicitly retain Bohr's affirmation: The yardsticks with which we measure distances and the clocks with which we measure time must have a high mass in order to remain unaffected by quantum conditions of uncertainty.

This essential remark had been overlooked by the founders of classical mechanics and by Einstein in his relativity theories; it will be necessary to closely reexamine its consequences in these various doctrines.

In one of his justly famous " little red books," Poincaré (1902) discusses the bases of mechanics:

> The English teach Mechanics as a experimental science. On the continent it is always presented more or less as a deductive science and *a priori*. The English are right, needless to say.... .
>
> On the other hand, if the principles of Mechanics have no other sources than experiments, they are therefore, only approximate and temporary.
>
> New experiments may lead us some day to modify or even abandon them.

Poincaré insists that the scientist should use, instead of arbitrary definitions, some conventions that are a résumé of empirical facts, and he states:

> Conventions, yes; arbitrary ones, no. They would be arbitrary if we lost sight of the experiments which led the founders of science to their adoption.

An identical point of view is expressed by Sommerfeld (1952).

Let us remember the warning about science's historical evolution and the negation of axioms or postulates given *a priori*.

To state an axiom is a logician's method, quite foreign to experimental science. Experimental science proceeds from empirical results, which may be codified, perhaps a little arbitrarily, in order to formulate working assumptions; these assumptions can be modified, if required, by experiment.

In any case it would be particularly naïve to believe in their universal validity. These hypotheses (called " laws " or even " principles ") apply only within certain limits, within a certain domain; the boundaries of this domain will be revealed to us by later experiments. (See Part I of SU.)

2. Classical Geometry, Kinetics, Classical Dynamics

In classical mechanics it is natural to start with *statics*, which represents a branch of geometry. Forces are treated as vectors, and neither movements nor masses are mentioned; applications are very numerous in the construction of bridges and in architecture. Then comes kinetics, where notions of time and space geometry intermingle: study of trajectories, description of motions (without trying to predict these motions).

Kinetics borrows the usual abstractions of classical geometry: points without dimension, infinitely thin curves, areas without thickness, etc. Curves are at one moment infinitely rigid, impossible to put out of shape, and then extremely flexible.

These methods (geometry plus time) enable us to define speeds, accelerations, changes in sets of coordinates, etc.

In classical mechanics " absolute " time is a fundamental assumption; in special relativity kinetics is modified by the introduction of relative times.

We note an essential point: Kinetics knows no notions of mass or force. Mass and force appear only in dynamics.

Laws of dynamics go back to Newton. For a brief but penetrating discussion, we refer the reader to the book on mechanics by Sommerfeld (1952, pp. 3–6).

Let us recall the three principles of dynamics:

1. Uniform and rectilinear motion in the absence of any outside force.
2. Definition of the quantity of motion (or momentum) that an outside force **f** may modify:

$$\mathbf{p} = m\mathbf{v}, \qquad \dot{\mathbf{p}} = \mathbf{f}$$

3. Equality between action and reaction.

The frequently quoted third principle is often given only lip service and later ignored.

The *laws of motion* are not sufficient to specify the precise trajectory of a material point. To complete the problem, we need to give (or rather *measure*) *initial conditions*. The mathematician generally forgets to mention this point in a discussion of the principles. We have emphasized in Chapters VI and VII of SU the facts that:

a. The laws of motion are reversible, and unsensitive to a reversal in the sign of time.
b. Initial conditions are irreversible because in reversing time, we change the sign of speed.

Consequently, *rational mechanics is, as a whole, irreversible.*

3. Frames of Reference in Classical Mechanics

Let us approach this thorny subject. It is customary to speak very little of it. Most authors keep kinetic and geometric definitions with all their unreal idealization, and get immediately busy with acceptable transformations: axes at rest, axes with a uniform motion, and relativity in classical mechanics or according to Einstein (see, for instance, Sommerfeld, 1952, pp. 9–16). In the course of the discussion one ignores the third principle of Newton, and one forgets as well the initial conditions.

Let us study this a little closer. To launch a projectile we need a machine: a crossbow, a musket, or a gun. This machine is tied to the frame of reference. It sustains a recoil and the frame of reference of no mass (ideal of geometricians) flies away! What we need is a motionless and stable frame of reference. It is necessary to provide it with an *infinite mass*. Thus, even in classical mechanics, we discover the importance of Bohr's remarks quoted in the first section.

In a discussion of classical relativity one compares two frames of reference, S_1 and S_2, with a certain relative speed v. The initial speeds of the projectile are v_1 and $(v_1 - v) = v_2$. The recoils sustained by the two frames at the moment of launching are different. The relative speed v is modified, therefore, if the two frames have finite masses, otherwise relativity breaks down. Instead of a launching machine, we can use a rocket; then the problems are more involved, since we have a projectile that splits into two fragments thrown right and left.

Let us go one step further in this discussion: We assumed that we were dealing with two frames of reference, S_1 and S_2, with a " given " relative speed v. Let us remember that nothing is " given " in a scientific investigation, *everything must be measured.* Only in a problem stated for an examination are certain quantities *given* by the teacher. How can we actually *measure the velocity v*? We build a laboratory on the first frame of reference S_1 and send some signals, optical signals for instance, to the S_2 frame. We measure the delay or the change of frequency in the Doppler effect, and we compute the velocity v from these observations. In the past century it was taken for granted that these measurements could be made without modifying anything in the moving systems Now we know that a photon $h\nu$ has a mass $h\nu/c^2$ and a momentum $h\nu/c$. While emitting a photon, our laboratory on S_1 feels the recoil, and when this photon strikes on system S_2 and is reflected back, there is a recoil on S_2.

When the photon strikes back on S_1, it perturbs again the state of motion of S_1. The relative velocity v of S_2 with respect to S_1 is modified by the measurements unless both frames of reference have very high masses M_1 and M_2. Theoretically the masses should be infinite, but it is sufficient to assume that they are enormously greater than the mass $h\nu/c^2$ of the photons used in the measurement.

The preceding discussion is only a special example of the Bohr–Heisenberg uncertainty relations and of the general remark that every experimental measurement means a perturbation. We shall find another example of these problems when we discuss the Doppler effect in Chapter 6.

Let us conclude: The usual statement of the relativity principle requires that frames of reference be extremely heavy.

Let us examine another role played by the third principle: A field of forces is defined, and its potential calculated (Sommerfeld, 1952, pp. 17–24). The force is given as a function of the coordinates x, y, z, and represents the *action* upon the projectile. But where is the *reaction*? Obviously, on the frame of reference that supplies the coordinates x, y, z. This frame must remain at rest: infinite mass!

Without clearly emphasizing this underlying assumption, Sommerfeld immediately speaks of the earth's gravitational field, the earth being motionless and without rotation—here is this practically infinite mass we were looking for.

Let us sum up: A frame of reference does not constitute a piece of unreal geometry anymore; it is a heavy laboratory, built on a rigid body of tremendous mass, as compared to masses in motion. Insufficient masses yield incomplete steadfastness—Here appear the effects of tides, with easily visible action and reaction.

We will be told: You are discovering the moon! No! We are only discovering that the moon should not be ignored in one chapter and correctly mentioned in another.

The use of accelerated frames of reference strengthens our arguments. What is the meaning of a frame S_2 in uniform rotation with respect to a motionless frame S_1? To give it a physical meaning, we have to see it as a very heavy wheel, a flywheel of high inertia, which carries away together the observer and the moving instrument under observation (both very light). If this condition is not fulfilled, any displacement of a mass m within the rotating frame modifies the moment of inertia of this frame and provokes a change in the speed of rotation ω. The flywheel must have an infinite moment of inertia so that we can consider ω as a constant when the observer and the moving instrument are arbitrarily displaced. The action (upon the moving instrument) is equal to the reaction (upon the frame of reference). The effect of this reaction can be ignored only if the mass of the frame is infinite.

These conditions are, moreover, realized in a laboratory on earth. The situation in an accelerated frame of reference is about the same as that observed in gravitation. A weighty object, observed on the earth, is placed into a field of gravitation which is in superposition with the effects of rotation. The action (upon the apple of Newton or upon our projectile) is equal to the reaction on earth.

4. Actions and Reactions in Relativity

In classical mechanics, all these effects are supposed to be transmitted instantaneously at any distance. In relativity, we require a transmission speed inferior to or equal to that of light c.

Here we must be very careful! It seems reasonable to assume that in a vacuum the propagation of gravitation follows a general, universal law. But what is to be thought of the propagation of a rotation? This is essentially a problem of the elasticity of the flywheel. When we set the flywheel in motion by applying a couple upon its axis, it first causes an elastic deformation in the steel of the flywheel. This deformation, first localized near the center, propagates progressively toward the periphery. It does not seem reasonable, nor justified, to imagine further any mechanism of universal propagation, as is the case for gravitation in a vacuum. Velocities of propagation for elastic waves that are very much inferior to c are the only ones to be considered. We cannot see much sense in talking of axes rotating in a vacuum. An interstellar neutral projectile does not feel the rotation of the earth, nor that of the sun, nor that of faraway stars.* Planets do not sense the rotation of the sun.

Einstein bases his general relativity on an *equivalence principle*: Gravitation and systems of rotating axes should be exactly of the same nature. This point of view does not seem justified. There exist similarities between these two sorts of phenomena. But there are also great differences.

When a landslide occurs inside the earth, it brings about elastic waves which propagate toward the surface of the planet, and we feel an earthquake. Up to this point, there is similarity with the setting in motion of a rotor, as described above.

But that is not all. The displacement of masses inside the earth brings about gravitation waves which propagate in space, outside our planet. The action of this perturbation can be felt on all objects and planets, even very distant ones.

This long distance action depends on a *universal constant* (Newton's constant), which does not play any role in rotating frames. The equivalence principle as stated by Einstein at the beginning of his theory is not, moreover, maintained later when the actual role of the curvature of the universe is discussed. One can wonder whether this equivalence actually constitutes a fundamental property, and whether its statement does not represent an extrapolation going far beyond experimental facts.

Einstein claims that the speed of gravitational waves equals the speed of light c. However, during the last fifty years, *no experimental* verification could be found. Gravitation may propagate with a speed much inferior to c, or even spread according to a law of diffusion or of heat propagation, nobody knows! (See Chapter 3.)

Concerning the propagation of actions and reactions, Einstein adopts Faraday's point of view of the *reality of fields*. The field propagating through empty space or through matter is composed of waves with finite velocities; in these waves one can find, beside the field, a second-order tensor (with four dimensions) which receives actions and reactions and transmits them step by step.

* It is fashionable to recall Mach's ideas about the origin of the notion of mass. These theories of Mach remained very vague and without experimental confirmation. We specify a " neutral " projectile because rotation of electric or magnetic bodies produces electromagnetic fields at a distance.

Einstein did specify the importance of the equality between these local actions and reactions ending in a propagation at a distance. In this manner, he says, the third principle of Newton is perfectly respected.

Is this so? How far are these actions going to propagate? They cannot last indefinitely. The waves which we imagine ending upon other material bodies will be reflected there, and so on, and so on, ..., and finally all will be lost at infinite distances! Here the question must be raised again: Which waves should we accept at infinite distances? On every boundary, some boundary conditions have to be specified, even if the boundary is at infinity. However, these boundary conditions have never been stated.

Levi-Civita (1937) and other authors select diverging waves (with retarded potentials) going even farther and carrying away their unwanted reactions. Einstein, Infeld, and Hoffmann (1938), on the contrary, use standing waves (superposed advanced and retarded ones) which adds up to imagining an immense mirror at infinity. This hypothesis is hard to accept.

The problem should be stated clearly and everybody should agree upon a boundary condition acceptable to all physicists.

What can we imagine? The problem of the behavior of gravitational waves looks like that raised for light or corpuscular waves. As far as light is concerned, we accept that it is lost at infinite distances, and that it spreads in the form of *retarded waves* (see Chapter VI, SU). We have no grounds for imagining that waves can come back from infinity. When one tries to specify these assumptions, one realizes how shaky they are. However, astronomers have never described anything which resembles a mirror at infinity. When gravitational waves are concerned, it seems reasonable (we do not say proven) to imagine similar conditions. This would make us select solutions of the type advanced by Levi-Civita, not those of Einstein. We would have to believe that forces of action and reaction are able to disappear at infinity. The third principle of Newton was strict in rational mechanics and gets blurred in relativity! The nonlinearity of gravitational waves complicates the situation, but for very great distances these waves become very weak and little by little linearity is restored.

The nature of the solutions at infinity is essential for the physicist and for the engineer. Is there or is there not emission of radiation? Can we dream of using Einstein's waves of gravitation to transmit signals? What is the speed of propagation of these waves? Could these transmission methods offer serious competition to the now overloaded electromagnetic radio? Many questions are left hanging.

5. Sets of Mathematical Coordinates or Physical Frames

Let us remind the reader of the origin of our discussions: In *geometry* or in kinetics, we use *sets of unreal coordinates*, supposedly infinitely rigid and without mass. One does not really talk of mass, since this notion appears only later, in *physical mechanics* and in dynamics. We have seen that at this second stage the frame of reference must be able to absorb reactions without moving. We therefore had to admit that this frame had an infinite mass; in order to establish this distinction, we propose *two different denominations*:

Sets of coordinates, rigid, no mass, in geometry
Frames of reference, infinite mass, in dynamics

Let us emphasize here that our definition of *heavy* frames of reference is in complete agreement with the selection of a clock based on the Mössbauer effect, a clock whose central atomic system is solidly embedded in a heavy crystal.

When reading Einstein's papers, one can readily see that he does not make this distinction and ascribes to sets of coordinates (without mass) properties that apply only to heavy frames of reference. But first let us discover in those very papers a premonition of the *frames of reference* (Einstein, 1911). In the second section of this paper, Einstein writes:

Let us consider two material systems S_1, S_2 provided with instruments of measurement.... .

Einstein does not specify that the mass of these material systems be very large, but he senses that a set of unreal coordinates is not sufficient and that one must imagine a material system, an actual laboratory of measurement.

In a later paper, which is the fundamental statement of general relativity, Einstein (1916) goes on, forgetting these precautions and making some surprising statements. In section 2:

We are able to " produce " a gravitational field merely by changing the system of co-ordinates.

In section 3:

> The general laws of nature are to be expressed by equations which hold good for all systems of co-ordinates, that is, are co-variant with respect to any substitutions whatever (generally co-variant).

At the end of this last sentence, we think should be specified: substitutions with a physical meaning, and representing an *actual operation in the Bridgman sense*. This is exactly the point when we disagree with Einstein.

Again in Section 3:

> In the general theory of relativity, space and time cannot be defined in such a way that differences of the spatial co-ordinates can be directly measured by the unit measuring-rod, or differences in the time co-ordinate by a standard clock.

This is a very dangerous statement, contrary to any experimental concept of science. We should be told how to effect these measurements. Otherwise the words " space " and " time " lose any physical meaning. We shall come back to this fundamental difficulty later.

Einstein's general sets of coordinates have been so popular as to be given the nickname " Einstein's mollusks." But can a physicist be forced to work under such conditions? It seems cruel to supply him only with rubber yardsticks and irregular clocks!

Finally in Section 4:

> According to the general theory of relativity, gravitation occupies an exceptional position with regard to other forces, particularly the electromagnetic forces, since the ten functions representing the gravitational field at the same time define the metrical properties of the space measured.

Einstein presents this statement as a property of nature, we would rather call it *Einstein's postulate*. All the effort of this author tends to reduce gravitation to geometry, and this at the price of a gulf, an actual breakup between gravitation and electromagnetism, through replacing the spatial potential of gravitation (Newton) with a tensorial potential of the second order, wrapping together gravitation and geometry. This is a genial mathematical work, but its application to physics remains open to discussion.

6. Fock's Assumption

One may hope to keep the theoretical method of Einstein, which still looks attractive, but it will be necessary to specify the definitions and set a limit to the conditions of application.

Moreover, the theory has already proven its excess of generality. Einstein himself declared that space and time cannot be related in a unique fashion to the results of measurements. No physicist will be satisfied by such a statement.

Let us consider a very remarkable book by the famous Russian scientist Fock (1964). This book contains a discussion of Einstein's ideas and rebuilds the usual Einstein theory from a very original viewpoint. Fock is able to obtain a practical solution of many difficulties. His most interesting result is the fact that it is *unreasonable* to keep the theory as *completely general as Einstein did*, and that some simple rules lead to a great simplification of the mathematical structure; at the same time he obtains a much better physical explanation of the practical meaning of the theory. He gets rid of unphysical generality and selects what he thinks is the best set of coordinates by assuming that the contracted Riemann–Cristoffel symbols Γ^α are zero (Fock, 1964, pp. 4, 193, 215):

$$\Gamma^\alpha = 0, \qquad \text{harmonic coordinates} \qquad (4.1)$$

These four additional conditions completely determine a " preferred " frame of reference, for which no correction is needed in the four-dimensional operation of wave propagation:

$$\Box \psi = 0 \qquad (4.2)$$

This means isotropic wave propagation with c playing the role of an absolute constant.

The theory developed by Fock requires careful examination. His method is certainly brilliant but it is not obvious whether his solution is the only possible one. He selects a certain class of frames of reference that simplify the solution, but perhaps there are other classes to be considered and compared with those considered by Fock. It also remains to be proven that Fock's selection of " preferred " frames corresponds to practical experimental conditions, especially with the modern definition of clocks (Chapter 3) and with the role played by the mass of frames of reference (this chapter).

Framingham State College
Framingham, Massachusetts

7. Schwarzschild's Problem

Some special examples may be helpful for a better understanding of the difficulties. Let us first consider the static problem of a particle at rest, with a field of spherical symmetry (see Pauli, 1958); we use coordinates x^1, x^2, x^3, and $x^4 = ct$ and obtain Schwarzschild's solution

$$ds^2 = (dx^1)^2 + (dx^2)^2 + (dx^3)^2$$
$$+ \frac{2m}{r^2(r - 2m)} \, [x^1 \, dx^1 + x^2 \, dx^2 + x^3 \, dx^3]^2$$
$$- \left(1 - \frac{2m}{r}\right)(dx^4)^2 \tag{4.3}$$

with

$$m = GM/c^2$$

and

$$r^2 = (x^1)^2 + (x^2)^2 + (x^3)^2$$

where M is the mass, G the gravitation constant of Newton, and m the reduced mass.

This solution becomes singular for a critical radius

$$r_0 = 2m \tag{4.4}$$

One should immediately note the possibility of deriving other mathematical solutions by suitable changes in the four coordinates. For instance, one may avoid the second group of terms in Eq. (4.3) and obtain isotropic space with

$$ds^2 = \left(1 + \frac{m}{2r}\right)^4 \, [(dx^1)^2 + (dx^2)^2 + (dx^3)^2]$$
$$- \left[\frac{1 - m/2r}{1 + m/2r}\right]^2 (dx^4)^2 \tag{4.5}$$

a new solution that collapses for

$$r_0 = m/2 \tag{4.6}$$

Both solutions behave similarly at infinity. Which one should we compare with experimental measurements? Should we select x^1,

x^2, x^3, and ct from Eq. (4.3) or from Eq. (4.5)? The question remains open. It is even worse than that: Any arbitrary change of coordinates can be applied and an infinite number of solutions obtained! Einstein's methods are much too general and do not yield any precise answer! Fock assumes that his condition (4.1) yields the frame of reference corresponding to actual physical observation. He obtains a third solution

$$ds^2 = \frac{r+m}{r-m} dr^2 + (r+m)^2 (d\theta^2 + \sin^2 \theta \, d\varphi^2) - c^2 \frac{r-m}{r+m} dt^2 \qquad (4.7)$$

This expression collapses for

$$r_0 = m \qquad (4.8)$$

The comparison of Eqs. (4.3), (4.5), and (4.7) clearly shows the trouble with Einstein's overgeneralization. And we may ask: Does Eq. (4.7) represent the last word? How can we prove that this solution actually corresponds to our length and time measurements in a laboratory at rest in a gravitational field? This cannot result from mathematical considerations but only from a careful discussion of actual experimental conditions. Such a detailed "operational analysis" according to Bridgman is absolutely needed, and it seems to be still missing. Looking at the preceding formulas, one feels that Eq. (4.3) looks awkward, and attempts at explaining it in physical terms are not too good. So we are left with Eq. (4.5) exhibiting local isotropy in space, or Eq. (4.7) characterized by local isotropy in wave propagation, a very troublesome situation indeed. Fortunately, in practice, reduced units result in an unbelievably small value for the critical radius r_0 so that the extremely small distances at which these catastrophes occur are practically unobservable.

We shall come back to this problem in Chapter 7 and reexamine it from a different point of view.

8. Quantum Theory versus Relativity

Two monumental theories were introduced in physics around 1900: Planck's quantum theory and Einstein's relativity theories. Now that more than sixty years have elapsed, we may compare their impacts on scientific thinking. *Quantum theory* is fundamental but constantly changing; its ideas are being subtly refined and readjusted almost every year to account for millions of new experimental results. One might count at least one hundred successive types of quantum theories. *Relativity* was built by Einstein into a most logical and rigid frame; special relativity was an enormous success, especially with the energy–mass relation. General relativity first seemed to be verified in three different types of experiments, of which two are seriously in doubt currently while the last one (red shift) checks very well with the latest experiments but can be explained by a much simpler theory (see Chapter 6). So we have to raise the question: General relativity is a splendid piece of mathematics, but what about its physical reality?

The general theory of gravitation, called general relativity (an unfortunate and misleading name, as emphasized by Fock) is based on the assumption that the gravity field propagates with the velocity of light waves c. No such effect has ever been observed although it may not be completely ruled out, as the very accurate observation of Weber (1967) shows.

If we investigate the matter further, we note that Einstein started his theory of general relativity with the aim of reducing gravitation and electromagnetism to space-time geometry. Hence, the obvious suggestion that both should have the same velocity c. Einstein succeeded in including gravitation in a four-dimensional geometry; but there was nothing to be done for electromagnetism. We now come back to the question stated at the beginning of Chapter 3. How does the gravitation field spread around?

a. As waves with a velocity $v_g \leqslant c$?
b. According to a diffusion equation?

There is absolutely no experimental answer and the question remains open.

The experimental " proofs " of general relativity were:

1. Deflection of light rays passing near the sun, observed during eclipses. These were very inaccurate experiments with individual errors of 100% and averaged errors of 30%. The theory is not safe because it assumes an ideal vacuum near the sun's surface, while we can observe very powerful explosions of matter and radiation from the sun.

2. The rotation of Mercury's perihelion. An apparently good check was proven largely accidental by Dicke (1967).

3. The red shift of spectral lines in a gravitation field. The Pound experiments brilliantly prove the result with 1% accuracy, but a very simple reasoning, using the mass $h\nu/c^2$ of a photon $h\nu$, is enough to make the prediction.

As a conclusion: There is no experimental check to support the very heavy mathematical structure of Einstein. All we find is another heavy structure of purely mathematical extensions, complements, or modifications without any more experimental evidence. To put it candidly, science fiction about cosmology—very interesting but hypothetical.

Altogether, we have no proof of the need for a curved universe (space plus time) and the physical meaning of this theory is very confusing.

REFERENCES

Bohr, N. (1958a). " Max Plank Festschrift," pp. 169–175, especially p. 171. Deut. Verlag. Wiss., Berlin.

Bohr, N. (1958b). " Atomic Physics and Human Knowledge," pp. 67–93. Wiley, New York.

Dicke, R. H. (1967). *Phys. Today* **20**, 55–70.

Einstein, A. (1911). *Ann. Physik* **35**, 898.

Einstein, A. (1916). *Ann. Physik* **49**, 769.

Einstein, A. (1923). " The Principle of Relativity." Dover, New York.

Einstein, A., Infield, L., and Hoffmann, B. (1938). *Ann. Math.* **39**, 65. See in particular the discussion on pp. 66 and 67 and the simple example discussed in paragraph 3 with its solution, Eq. (3.18).

Fock, V. (1964). " The Theory of Space, Time and Gravitation." Pergamon Press, Oxford.

Levi-Civita (1937). *Am. J. Math.* **59**, 225.

Pauli, W. (1958). " Theory of Relativity," p. 166. Pergamon Press, Oxford.

Poincaré, H. (1902). " La Science et l'hypothèse," pp. 110–148. Flammarion, Paris.

Post, R. J. (1967). *Rev. Mod. Phys.* **39**, 475–493.

Sommerfeld, A. (1952). " Mechanics," Vol. 1. Academic Press, New York.

Weber, J. (1967). *Phys. Rev. Letters* **18**, 498–501.

Chapter 5

**Special Relativity
Doppler Effect**

1. A Reappraisal of Fundamental Assumptions

In the preceding chapters we analyzed a variety of experimental evidences that led us to a complete reappraisal of many basic assumptions in theoretical physics. We emphasized in Chapter 3 the fundamental importance of the Mössbauer effect and of atomic clocks, both experimental developments that Einstein could not foresee and that enable us at present to give a very precise empirical (and quantum theoretical) definition of *ideal clocks*. This modern definition of a clock must precede any discussion of what an ideal clock actually does. This is not mere guesswork, but logical reasoning according to Bridgman's operational method, or, better said, according to the traditional method of experimental science, which was responsible for the enormous success of modern science and definitely created natural philosophy as distinct from metaphysics.

This line of discussion was developed in Chapter 4, where a number of classical problems of theoretical mechanics were discussed and it was finally stated that a sharp distinction should be made between geometry and physics. The sets of coordinates in geometry should not be confused with the frames of reference of physics. The notion of *mass* is unknown in geometry and cannot be defined without physical experiments. The frames of reference of physics are schematizations of actual physical laboratories, some solid and heavy structures containing a variety of measuring instruments.

This general viewpoint is in complete agreement with the definition of an atomic or Mössbauer clock specified in Chapter 4. The Mössbauer clock has an exceptional character of using an atom embedded in a heavy crystal structure that itself is resting in a very heavy, physical frame of reference.

Our viewpoint is, however, in contradiction with Einstein's assumptions when he was dreaming of reducing physics to a branch of geometry and taking for granted the possibility of doing this with the help of a Gauss–Riemann geometry. Einstein's dream comes very close to real physics, but it is in contradiction with the definition of actual atomic clocks, and one should not forget that these clocks represent the most remarkable measuring instrument in physics, with an accuracy reaching 10^{13} for atomic clocks or 10^{16} for future Mössbauer clocks.

The restricted relativity of Einstein represents an extraordinary achievement, but we shall see that the so-called general relativity may only be considered as an approximation, and certainly needs a thorough revision.

2. Recoil Problem for Atoms

Let us consider an atom and assume that it remains *at rest* in a certain frame of reference. This atom may have some energy levels, E_1 and E_2 say, and is able to emit a frequency $h\nu_0$, where ν_0 is the unperturbed frequency, according to Bohr's condition

$$E_1 - E_2 = h\nu_0 \qquad (5.1)$$

We also remember the famous mass–energy relations:

$$E_1 = M_1 c^2, \qquad E_2 = M_2 c^2 \qquad (5.2)$$

where M_1 and M_2 represent the total mass of the atom (including the rest-mass of the nucleus, etc.) on its energy levels, E_1 or E_2. Condition (5.2) includes the fact that $M_1 \mathbf{v}_1$ and $M_2 \mathbf{v}_2$ do represent the momenta.

Here we sound a note of caution. The atom initially at rest was falling from energy E_1 to E_2 and emitting a photon $h\nu$; this photon, according to Einstein, carries a momentum $h\nu/c$; hence we have a recoil on the atom (mass M_2) which takes a velocity v.

Conservation of momentum requires

$$-\frac{h\nu}{c} = M_2 v = E_2 \frac{v}{c^2} = \frac{1}{c} E_2 \beta \qquad \text{with} \quad \beta = \frac{v}{c} \ll 1 \qquad (5.3)$$

and by this recoil the atom obtains a kinetic energy

$$E_{\text{kin}} = \tfrac{1}{2} M_2 v^2 = E_2 v^2 / 2c^2 = \tfrac{1}{2} E_2 \beta^2 \qquad (5.4)$$

This energy must appear in the energy balance and relation (5.1) must be replaced by

$$E_1 - E_2 = h\nu + \tfrac{1}{2} E_2 \beta^2 \qquad (5.5)$$

with new frequency ν perturbed by classical recoil.

This classical result is only a first approximation for $v \ll c$. We may compute a better approximation with the help of relativistic mechanics, replacing Eqs. (5.3) and (5.4) by

$$E_{\text{kin}} = E_2 \left[\frac{1}{(1-\beta^2)^{1/2}} - 1 \right]$$

$$-\frac{h\nu}{c} = \frac{E_2}{(1-\beta^2)^{1/2}} \frac{v}{c^2} \qquad (5.6)$$

The second equation reads

$$\frac{E_2 \beta}{(1-\beta^2)^{1/2}} = -h\nu \qquad \text{or} \qquad \frac{\beta}{(1-\beta^2)^{1/2}} = -\frac{h\nu}{E_2} = \alpha \qquad (5.7)$$

This is easily solved

$$\frac{\beta^2}{1-\beta^2} = \alpha^2, \qquad \frac{1}{1-\beta^2} = 1 + \alpha^2 \qquad (5.8)$$

hence,

$$E_{\text{kin}} = E_2 [(1+\alpha^2)^{1/2} - 1] = h\nu - h\nu_0$$

This relation reduces to (5.5) when β is very small

$$E_2 \gg h\nu, \qquad \alpha \ll 1, \quad \beta \ll 1 \qquad (5.9)$$

What we want to make clear is that the actual frequency ν differs from the " frequency of an atom at rest " by a term that can become negligible only when the energy E_2 of the atom is very much greater than $h\nu$.

A frame of reference at rest requires a large rest energy, hence a heavy mass. This is a typical example of the general result of Chapter 4. It also checks with experimental results that lead to the choice of cesium (the heaviest of alkaline metals) for astronomical standard clocks, or to the use of Mössbauer rays in Pound's experiments.

3. Doppler Effect

A careful discussion of the Doppler effect leads to similar conclusions. Here we refer to a paper by Schrödinger (1922) and explanations given by Sommerfeld (1954).

An atom on energy level E_1 comes to point 0 (Fig. 5.1) with incident velocity v_1 at angle θ_1 and emits a quantum $h\nu$ in the x

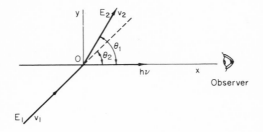

FIG. 5.1

direction. After this emission the atom retains an energy E_2 and a velocity v_2 at angle θ_2. We write the three equations for energy and momentum conservation in the x and y directions:

$$\frac{E_1}{(1-\beta_1^2)^{1/2}} = \frac{E_2}{(1-\beta_2^2)^{1/2}} + h\nu, \quad \beta_i = v_i/c, \quad i = 1, 2 \quad (5.10)$$

$$\frac{E_1}{(1-\beta_1^2)^{1/2}} \frac{v_1 \cos \theta_1}{c^2} = \frac{E_2}{(1-\beta_2^2)^{1/2}} \frac{v_2 \cos \theta_2}{c^2} + \frac{h\nu}{c} \quad (5.11)$$

$$\frac{E_1}{(1-\beta_1^2)^{1/2}} \frac{v_1 \sin \theta_1}{c^2} = \frac{E_2}{(1-\beta_2^2)^{1/2}} \frac{v_2 \sin \theta_2}{c^2} \quad (5.12)$$

Combining (5.10) and (5.11) we obtain

$$\frac{E_1}{(1 - \beta_1{}^2)^{1/2}} (c - v_1 \cos \theta_1) = \frac{E_2}{(1 - \beta_1{}^2)^{1/2}} (c - v_2 \cos \theta_2) \quad (5.13)$$

Let us introduce the notations

$$\varphi_i = \frac{c - v_i \cos \theta_i}{(c^2 - v_i{}^2)^{1/2}}, \qquad i = 1, 2$$

$$\psi_i = \frac{v_i \sin \theta_i}{(c^2 - v_i{}^2)^{1/2}}, \qquad i = 1, 2 \quad (5.14)$$

We obtain from Eqs. (5.12) and (5.13)

$$E_1 \varphi_1 = E_2 \varphi_2 = \alpha \quad \text{and} \quad E_1 \psi_1 = E_2 \psi_2 = \gamma \quad (5.15)$$

where α and γ are constants. Next, we have the identity

$$(c^2 - v_i{}^2)^{1/2} = \frac{2c\varphi_i}{1 + \varphi_i{}^2 + \psi_i{}^2} \quad (5.16)$$

and our first equation (5.10) reduces to

$$h\nu = E_1 \frac{1 + \varphi_1{}^2 + \psi_1{}^2}{2\varphi_1} - E_2 \frac{1 + \varphi_2{}^2 + \psi_2{}^2}{2\varphi_2}$$

$$= \frac{1}{2\alpha} (E_1{}^2 + \alpha^2 + \gamma^2) - \frac{1}{2\alpha} (E_2{}^2 + \alpha^2 + \gamma^2)$$

$$= \frac{1}{2\alpha} (E_1{}^2 - E_2{}^2) \quad (5.17)$$

with the help of Eqs. (5.15). Hence,

$$h\nu = \frac{E_1 + E_2}{2\alpha} (E_1 - E_2) = \frac{E_0}{\alpha} h\nu_0, \qquad E_0 = \frac{E_1 + E_2}{2} \quad (5.18)$$

when E_0 is the average of E_1 and E_2, while ν_0 represents the unperturbed frequency (5.1). We want to compare this result with the *classical relativistic Doppler effect* ν_D corresponding to an arbitrary velocity v_0:

$$\nu_0 = \frac{c - v_0 \cos \theta_0}{(c^2 - v_0{}^2)^{1/2}} \nu_D = \varphi_0 \nu_D \quad (5.19)$$

with the notation of (5.14). The difference between our result (5.18) and the standard Doppler formula (5.19) comes from the fact that we have two energy levels E_1 and E_2, two velocities v_1 and v_2 and two angles θ_1 and θ_2 while the classical formula contains only one E_0, one v_0 and one θ_0. The *classical Doppler formula* (5.19) can be considered only as a *first approximation* provided

$$v_1 \approx v_2 \approx v_0, \qquad \theta_1 \approx \theta_2 \approx \theta_0, \qquad E_1 \approx E_2 \approx E_0$$

these conditions requiring obviously a very small recoil impulse. Hence,

$$E_2 \gg h\nu \qquad\qquad (5.20)$$

which is again the same condition as in (5.9).

Classical relativistic formulas are valid only when the total energy of the atom is very large compared with the change occurring during quantum transition. This corresponds exactly to what we stated in Chapter 3. *A physical frame of reference must be very heavy.*

4. Discussion of the Actual Quantized Doppler Effect

Let us reexamine the experiment sketched in Fig. 5.1. We have a fixed frame of reference $0xy$, at rest, and our observer (or observing apparatus) is at some distance to the right in the x direction. We observe radiation emitted from a moving atom (proper energy level E_1), velocity v_1, angle θ_1, and we know that this radiation is emitted when the atom drops to another proper energy level E_2. We have the initial data

$$E_1, \qquad E_2, \qquad v_1, \qquad \theta_1 \qquad\qquad (5.21)$$

After the emission the atom obtains a different velocity (v_2, θ_2) but we do not observe this velocity directly. What we observe is the frequency ν of the radiation emitted in the x direction, and we compare it to the frequency ν_0 that would have been emitted by an atom at rest with unperturbed frequency ν_0

$$h\nu_0 = E_1 - E_2 \qquad\qquad (5.22)$$

Our formula (5.18), combined with (5.15) and (5.14) yields the quantized Doppler effect:

$$\frac{\nu_0}{\nu} = \frac{2\alpha}{E_1 + E_2} = \frac{2\alpha}{E_1(1 + E_2/E_1)}$$

$$= \frac{2\varphi_1}{1 + (E_2/E_1)}$$

$$= \left[\frac{2}{1 + E_2/E_1}\right] \frac{c - v_1 \cos\theta_1}{(c^2 - v_1^2)^{1/2}} \tag{5.23}$$

We managed to keep in this final formula, only the initial data and the frequencies ν_0 (unperturbed) and ν (observed).

Comparing this final result with the classical Doppler effect (5.19), we note that the quantized Doppler effect contains the bracketed additional factor

$$1 \leqslant \left(\frac{2}{1 + E_2/E_1}\right) \leqslant 2 \tag{5.24}$$

This factor is nearly unity when E_2 is very close to E_1, and this corresponds to the standard Doppler effect discussed at the end of the preceding section. The same factor may go up to 2 when E_2 becomes smaller and smaller. The limit 2 is obtained when E_2 is zero, which means total disintegration of the atom into two photons, one observed along x and the second one escaping with $v_2 = c$ in the direction θ_2.

5. A Correct Statement of the Principle of Relativity

We sketched in Chapter 1 the historic development of the principle of relativity. The point of departure was found in classical mechanics, where it was discovered that all the laws of motion were exactly similar within a frame of reference at rest or within a frame moving with a given constant velocity v. After this first step (emphasized by Poincaré) came the discovery that relative motions of translation could not be observed in electromagnetism also.

Right at the beginning we must clearly state a *difficulty* which seems to have been *overlooked by most authors*. It results from the discussion of Chapter 4.

What does it mean to speak of a frame of reference moving with a given constant velocity v, with respect to a frame at rest?

This statement does not make sense unless both systems (the one at rest and the other one in motion) are very much heavier than anything observed. It is meaningless to speak of a " given " velocity. When an astronomer discovers a new moving body in space, he does not initially know the velocity. In order to *measure the velocity*, he must proceed with experiments. He may send some bullets to the unknown object, have them reflected back to him and observe the time intervals in pulses; in order not to disturb too much the moving object, he will choose the lightest known bullets: light *photons*. But even photons do have a mass; when emitted they push back the " frame at rest " by recoil and when reflected they push away the frame in " uniform motion." The uniform relative motion is perturbed, and our definition is meaningless. We rediscover here the well-known rule: *Any experiment means a perturbation*. The perturbation may only become negligible if the masses of both systems are really huge, very much greater than those of our photons.

This is again the problem discussed in Section (4.3) about the very *definition of a frame of reference in physics*. The Schrödinger discussion of Section 5.4 about the Doppler effect means exactly the same thing. If our atom has a small initial energy E_1, the final energy E_2 is much smaller than E_1 and the factor $2E_1/(E_1 + E_2)$ of Eq. (5.24) may differ very much from unity. This factor corresponds directly to the recoil effects, and the emission of $h\nu$ modifies the velocity $(v_2 \neq v_1)$. The frame of reference cannot maintain a constant velocity.

Frames of reference, either at rest or moving with a constant velocity, represent an *idealization* that can be used only for very heavy systems.

This sort of idealization was taken for granted in past centuries, when everybody assumed that " just looking " at a moving object could not perturb its motion; but we know that this assumption was wrong and cannot be maintained today.

In the preceding discussions we used a discovery of Einstein; the *light quantum*, or *photon*. This was the discovery for which a Nobel prize was granted to Einstein *because of so much experimental evidence* about its physical reality. Nevertheless, Einstein never liked his photon as tenderly as his beloved relativity. The photon was a natural child, a bastard born out of wedlock; Einstein remained a strong believer in differential equations in a continuous medium. Discontinuities and quanta seemed to him unnatural.

It is, however, with quantum conditions and photons that (much to our surprise) we discover the fundamental laws of physics and the modern definition of clocks!

6. How Does a Moving Clock Behave?

What we said about an atom can actually be observed with a moving clock, since a modern clock is nothing but a laser system synchronized on one atomic frequency. Einstein could not have foreseen this experimental definition of an ideal clock; he could not have imagined how the clock would appear to a moving observer, nor how a moving clock would behave when observed from a frame of reference at rest.

The clock is a piece of apparatus emitting a definite frequency ν_0 in a frame of reference where it stays at rest. It represents a standard frequency. When the clock it moving with a constant velocity v, we may observe a whole spectrum of frequencies depending upon the direction of the velocity v, and the whole result is described by our formula (5.23) for the *quantized Doppler effect*.

Instead of speaking of a variety of *frequencies emitted* by Doppler effect, Einstein spoke of a modified scale of time and this led to all sorts of paradoxes.

The ideal clock should be very heavy (Mössbauer effect, for instance) in order to get rid of the correction in E_2/E_1, but this correction is easy enough to apply when it may be needed. It is curious to note that Einstein paid so much attention to the factor $(1 - \beta^2)^{1/2}$ and practically discarded as uninteresting the whole structure of the Doppler effect. Actually, this effect is a whole that cannot be split to pieces. How did Einstein happen to come to such conclusions? He started from the Lorentz transformation; this transformation is usually written for a simplified problem that we shall discuss in Section 5.7. The Lorentz transformation suggests a length contraction along x and a similar time contraction.

From the Lorentz transformation, Einstein (1905) computes the Doppler effect by using an oblique velocity and obtains formula (5.19), but all the discussions of time scales and lengths measurements are based upon the Lorentz formula, where an oblique velocity is never considered.

There is, in addition to this, a curious coincidence that was noted by Schrödinger (1922): If one wants to obtain the correct Doppler effect by the computation of Section 5.3, one must include the momentum of the photon. Without this consideration we would be left with a single equation (5.10) that contains only the $(1 - \beta^2)^{1/2}$ effect and that corresponds to an arbitrary simplification similar to that of Einstein.

In the discussions of the Doppler effect, we had the $(1 - \beta^2)^{1/2}$ terms appearing directly in the three fundamental formulas (5.10–12) and these terms simply revealed the existence of kinetic energy for both initial and final stages. The whole computation was based only on the quantum condition (5.1), the mass–energy relation (5.2) and the principles of conservation of energy and momentum.

The main point is that we have to use a *model of a clock that is very different* from the one Einstein had in mind. He visualized a clock as a sort of radar apparatus emitting short signals and measuring time intervals between such sharp signals. We now have clocks emitting continuous oscillations of a given frequency, but these clocks are not built for the emission of sharp signals. We look at the clock model from a very different viewpoint, and this description of the clock brings the whole Doppler effect into the foreground, while the Lorentz transformation just means a mathematical tool. The interest is shifted from mathematics to physical facts. This is stressed also by the remark that *frames of reference must be heavy,* and that we must not talk of accelerating or decelerating them arbitrarily. Let us think of the laboratory at rest as a railway station while the moving frame of reference is a heavy train. This provides a good representation of what happens when velocity remains constant, but *we do not know and should not guess what may happen to an accelerated clock.* Descartes introduced a wonderful method when he invented *sets of coordinates* but the method is *a terribly artificial one* because it requires an origin of coordinates and an origin of time, for which we have no definition. Hence, all the results of importance must be independent of the choice of the origin. As soon as we speak of sets of coordinates we must state a *principle of invariance from the choice of the origin.* The only quantities that matter are *distances between points* and *intervals of time.* When we choose to center our attention on the Doppler effect, we avoid all these unnecessary complications. The *Doppler effect* corresponds to the actual fundamental observation.

7. A New Approach to Special Relativity

All authors writing about relativity follow the same road:

Michelson experiments → Lorentz transformation
$$\rightarrow \text{Einstein's Theory} \qquad (5.25)$$

Traveling on this highway, the teacher misses many important viewpoints, which may be discovered if we travel more leisurely on another road.

Along the many experimental proofs of special relativity, we may select those of greatest importance, and use these experimental results as starting points. We intend to use the following path:

Mass–energy relation → Atomic clock → Doppler effect
$$\rightarrow \text{Lorentz transformation} \qquad (5.25')$$

This last step requires a *special assumption* which was clearly stated by Einstein, but which is usually overlooked by modern writers, as if it were obvious; this is, however, not the case and a special discussion is required.

The mass–energy relation, proved by atomic bombs, is summarized by the equations

$$E = Mc^2$$

$$M = \frac{M_0}{(1 - v^2/c^2)^{1/2}} \qquad (5.2')$$

$$\mathbf{p} = M\mathbf{v}$$

for a particle of rest mass M_0, and

$$E = h\nu = Mc^2, \qquad p = \frac{h\nu}{c} \qquad (5.3')$$

for photons of zero rest mass. The atomic clock discussed in Chapter 3 rests on Bohr's second condition

$$\Delta E = h\nu, \qquad \text{period} \quad \tau = \nu^{-1} \qquad (5.1')$$

These definitions, completed by the principles of conservation of energy and conservation of momentum are all we need for computation of the Doppler effect (Sections 5.3 and 5.4). The atomic clock defines a single frequency ν_0 in the frame of reference where the clock is at rest. The word " rest " implies a very high mass for the clock and the frame (Sections 5.5 and 5.6).

When the clock is observed from a heavy moving frame of reference, it is seen to emit frequencies ν that depend on the direction θ of observation [Eq. (5.23)]. Let ν_0 be the frequency in the frame at rest and τ_0 the period while ν is the frequency and τ the period observed in the moving frame at angle θ_1; formula (5.23) now reads

$$\frac{\tau}{\tau_0} = \frac{\nu_0}{\nu} = \frac{c - v_1 \cos \theta_1}{(c^2 - v_1{}^2)^{1/2}} \qquad (5.26)$$

since periods τ and τ_0 represent ν^{-1} and $\nu_0{}^{-1}$, respectively. Thus far we strictly follow the Doppler computation.

In order to rejoin the *Lorentz formulas*, we must remember the history of the subject. Lorentz was thinking in terms of the Michelson experiments, where the complete Doppler effect could not be observed. In Michelson's experiments light rays were always traveling back and forth in both directions, all along the circuit of light beams. This means that the velocity v_1 was always associated with a $\pm$ sign. Michelson could observe nothing more than the average $\pm v_1$. Introducing this average in Eq. (5.26) we get rid of the $v_1 \cos \theta_1$ term:

$$\left(\frac{\tau}{\tau_0}\right)_{av} = \left(\frac{\nu_0}{\nu}\right)_{av} = \frac{c}{(c^2 - v_1{}^2)^{1/2}} \qquad (5.27)$$

and this is the transformation of time in the Lorentz formula.

Experiments of Joos, using a Michelson device, were about five times more accurate than those of Miller. New experiments by Townes (1958) and co-workers used *two masers emitting beams propagating in opposite directions*. The orientation of the apparatus with respect to the motion of the earth was modified and the whole device gave results fifty times more accurate than those of Joos. Modern lasers can yield much higher accuracy.

8. The Lorentz Transformation

The Lorentz transformation requires an averaging with light beams propagating in opposite directions. This means that the velocity of light is measured for signals traveling back and forth over a certain distance, and this is required logically by Einstein's remark that time coincidence is impossible to define between two points at a distance. Only *time and space* coincidence have a physical meaning and can be observed. In addition, Einstein assumes *space symmetry* in a frame of reference at rest, and *also* in a frame of reference in uniform motion, since we cannot detect the motion by observations made within the moving frame of reference.

This situation is clearly specified in discussions using the following well-known model: The *frame of reference at rest* is supposed to be a long railway station, extending for quite a distance along the tracks. The *moving frame of reference* is a heavy train moving on the tracks. The station is equipped with fixed clocks all along the tracks, and these clocks have been synchronized by signals sent back and forth from a central clock in the station. These signals are assumed to be propagating with velocity c in the frame at rest. The train is equipped with clocks in all its cars, these clocks being synchronized with a central clock on the engine. Here again back and forth signals have been used, *assuming a velocity c both ways, with respect to the train.* This assumption (emphasized by Einstein) is based on the fact that no experiment made on the train can detect its constant velocity. There is complete symmetry between station and train. The station master looks at the clocks in successive cars as they run through the station, and he sees these clocks going slow (because of the method by which they were synchronized). The same is true for the engineer looking down at the successive clocks he sees along the track; both situations are exactly symmetrical.

Altogether, we find complete agreement with the theory of special relativity, although we attacked the problem from a completely new point of departure.

We have to emphasize the very important role played by *Einstein's rule for synchronization of clocks* and setting them right in each frame of reference; this *rule is arbitrary and even metaphysical.* It can neither be proved nor disproved by experiments, it *assumes* that signals propagating east to west or west to east have equal velocities, while Michelson's experiments only measure the average of these two velocities. The sudden and unverifiable assumption is obvious. Our discussion of the complete problem with Doppler effect shows that the actual physical facts do not prove Einstein's assumption, while this assumption is required for the Lorentz transformation.

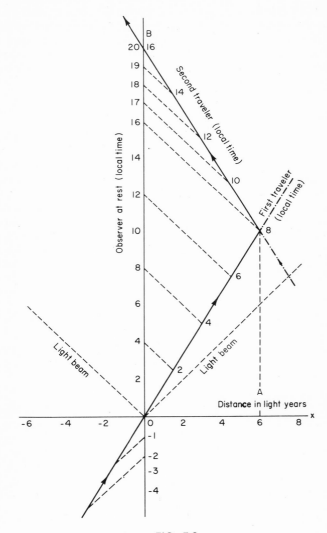

FIG. 5.2

Einstein's results are correct but the *Lorentz transformation* is a *mathematical, unobservable* tool—very useful, but *definitely not physical*. Similarly, the ds^2 of Minkowski is a most interesting expression to consider but *definitely not physical*; in both cases the synchronization rule is needed and not proven, although it cannot be disproved either.

9. The Problem of Traveling Twins

This is a classical problem for endless discussions. [Many of these discussions may be found in a collection of selected reprints of the American Association of Physics Teachers (1963).] One twin stays quietly on his home at O. The other one travels with great speed to a distant point A and immediately comes back home; when he is welcomed by his brother, they discover that the traveler is distinctly younger than the fellow at home. There is a moral to the fable, but let us omit it and discuss the facts. A numerical example simplifies the graphical discussion. Let us assume $v = 0.6c$, hence

$$(1 - v^2/c^2)^{1/2} = 0.8$$

Then the Doppler factors are

$$\underset{\rightarrow}{K} = \left(\frac{c - v}{c + v}\right)^{1/2} = \tfrac{1}{2} \qquad \text{on the way out}$$
$$\underset{\leftarrow}{K} = \left(\frac{c + v}{c - v}\right)^{1/2} = 2 \qquad \text{on the way back} \tag{5.28}$$

The distance OA is six light years.

In Fig. (5.2) we take the abscissa for the distance x (in light years) and the ordinate for time t. The lazy twin, staying at home, is represented by points along the vertical axis, and his rest time is indicated. Dotted lines represent signals emitted by the traveling twin every two hours (on his own clock). His trip away takes eight years, then he turns back and comes back home after sixteen years (on his clock) to meet his lazy twin whose clock shows twenty years!

Signals emitted by the traveler after
  2, 4, 6, 8, 10, 12, 14, 16 local years
are received by the twin at rest at
  4, 8, 12, 16, 17, 18, 19, 20 years.

Einstein's formula gives the total change (sixteen to twenty) but the signals in dotted lines show the Doppler effect, and there is no possibility of confusing the twin at rest with his traveling brother. Of course, the traveler is characterized by the fact that he supports a strong acceleration at the moment he reaches point *B*. We can choose not to speak of this disturbing acceleration if we use triplets instead of twins: one of the triplet stays at home, the second one travels away and never comes back, and the third one travels back and meets the second one at *B*.

Here there is no question of acceleration. We simply have first and second brothers comparing their watches at the start when they meet at home, second and third brothers comparing their watches when they meet at *A*, third and first brothers comparing their watches when they meet at home at the end.

The dissymmetry between the one at rest and the two others traveling in opposite directions is now obvious.

The Lorentz formula is correct for the final result, since both directions of travel (forth and back) are used, but the consideration of Doppler effects on signals emitted by the travelers reveals the complexity of the problem. The description with the triplets is simpler, since there is no question about acceleration, but only a question of comparing clocks at the moment when they happen to be together at the same point, an operation of true physical meaning.

All our discussions, although starting from an unconventional point of departure, completely agree with Einstein's special relativity; we only emphasize the condition that every part of the circuit of *light beams must be traveled in both directions* if we want to eliminate the details of the Doppler effect and keep only the Lorentz transformation. Also of importance is the remark that *frames of reference must be heavy* (Chapter 4). It may suggest that we must be prepared for troubles when we apply the idea of relativity to very light particles. There may be a need for corrections, as we discovered in the case of the Doppler effect. Nevertheless, this theory represents Einstein's most remarkable achievement, and the famous mass–energy relation is fundamental throughout physics.

REFERENCES

American Association of Physics Teachers (1963). " Special Relativity Theory, Selected Reprints." American Institute of Physics, New York. See especially the articles by Scott, p. 80 (from *Am. J. Phys.* **27** (11), 580); and Frisch, p. 89 (from *Contemp. Phys.* **2** (10), 61; **3** (2), 194).

Schrödinger, E. (1922). *Physik. Z.* **23**, 301–303.

Sommerfeld, A. (1954). " Optics," p. 85. Academic Press, New York.

Chapter 6   **Relativity and Gravitation**

1. The Mystery of Gravitation

Special relativity theory has to be reconciled with the theory of gravitation. Gravity was assumed by Newton to be propagated instantly at any distance, an assumption that appeared very risky at the time: How can one imagine no delay for actions transmitted across the fantastic distances of the universe. Einstein assumed a gravitation velocity equal to that of light

$$v_g = c \tag{6.1}$$

but we noticed in Chapter 3 that the only thing we can say is that v_g is smaller than or equal to c. As a matter of fact, it is really remarkable that we still know nothing experimentally about the propagation of gravity, despite so many skillful experimental attempts since the beginning of this century!

We have to rely on a few elementary observations, the first of them being the law of Galileo: In a vacuum all bodies fall with equal acceleration. Eötvös checked this law with great accuracy and it can best be stated by the relation

$$M_{\text{gravific}} = M_{\text{inertial}} \tag{6.2}$$

The gravific mass always equals inertial mass. Another way to state this is to notice that, in the vicinity of a given point in space and time, the gravitational field can be imitated or compensated by a field of acceleration. This is what Einstein calls the " principle of equivalence."

Einstein's ideas on equivalence are well known. The best discussion, in our opinion, was given by Fock (1964) starting from Eq. (5.2). The explanations given in Fock's book are really illuminating and are recommended to the reader (see especially the Introduction, and Chapters V–VII). Fock emphasizes a few points of great importance:

a. It is not sufficient to study space and time locally in infinitely small regions, just as it is not enough in classical mechanics to state local equations of motion. One must also specify the boundary conditions (or the initial conditions in classical mechanics), otherwise the problem is not completely defined. Local conditions and boundary conditions are inextricably interconnected; even when the boundary is at infinite distance the boundary conditions are absolutely necessary, and should never be overlooked.

b. Einstein assumed the necessity of *not choosing any preferred systems of coordinates*, and this led to an overgeneralization that was very confusing.

If space and time are uniform at infinity, it is possible to introduce a *preferred* system of coordinates (defined up to a Lorentz transformation) that Fock calls *harmonic coordinates*. These coordinates are characterized by a condition containing the four contracted Riemann–Christoffel symbols, which are equated to zero

$$\Gamma^\nu = g^{\alpha\beta}\Gamma^\nu_{\alpha\beta} = 0 \tag{6.3}$$

These conditions do not introduce any essential limitations on the solution, but they narrow down the generality. This point had first been noted by de Donder (1921) and Lanczos (1922). We already quoted Fock in Eqs. (4.1) and (4.2) and we noted that in order to justify physically this choice of preferred coordinates, it would be necessary to prove that the x, y, z, and t thus defined correspond to the quantities actually measured in a physical laboratory. Fock's mathematical proof of a very general simplification is most interesting, but *it remains to be shown that his assumption is compatible with the role of mass in a frame of reference* (Chapter 4) and with the modern definition of a cesium clock (Chapter 3) or a Mössbauer frequency standard.

In the present uncertainty about experimental laws of gravity, we are inclined to trust Fock's presentation of the theory rather than Einstein's general relativity, which appears too general and too far from physical reality.

We completely agree with Fock on the impossibility of splitting the problem into two separate parts, as is usually done by mathematicians. There is absolutely no reason for discussing separately the local conditions (equations of motion or wave equations) and relegating to a second place the boundary conditions (or initial conditions) which are supposed to be " given " arbitrarily for each problem. Nothing can be taken for granted and nothing is ever given free in experimental science.

In the following sections we intend to base our discussion on the definition of the atomic clock (Bohr's second condition) and the general existence of energy levels over all chapters of physics. The lack of experimental data on gravity propagation leaves us in the dark, groping for our way, and we shall try to constantly check our deductions against empirical facts.

First, we do not want to make any arbitrary assumptions about gravity propagation, and we shall restrain ourselves to *steady state* conditions. We continue to assume Euclidian space and want to discuss the behavior of an atomic clock in a static gravity field.

Following our discussion of the Doppler effect in Chapter 3, we shall emphasize the most important role played by the photon.

2. An Ideal Atomic or Mössbauer Clock and the Gravity Red Shift

Let us investigate the definition of an ideal clock and the problem of the gravity red shift from a strictly experimental and operational point of view.

Our assumptions are explained in Fig. 6.1: A spherical body of mass M, at rest, yields a gravitation potential V at a distance r. This potential is zero at infinity and takes a negative value V_a on the surface of the sphere, where we have (at $r = a$) an experimental *laboratory at rest*. We compare an atomic clock located at infinity (potential and field both zero) with a similar clock *at rest* at point A under gravitational potential V_a and a force f_a

$$V_a < 0, \qquad f_a = -m\frac{\partial V_a}{\partial a} \qquad (6.4)$$

where m is the mass of the atom. The heavy sphere of mass M

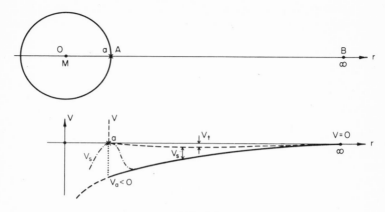

FIG. 6.1

unambiguously defines our frame of reference at rest, where we may build a laboratory at rest at point A. All conditions required in Chapter 4 are fulfilled.

At zero potential and velocity zero, the atom has two energy levels E_{1_0} and E_{2_0}, whose difference yields a radiation frequency:

$$E_{1_0} - E_{2_0} = h\nu_0 = m_1c^2 - m_2c^2 \qquad (6.5)$$

where m_1 and m_2 are the masses of the atom on both levels. Let us now watch the *atom* (on level 1) fall from infinity to a, where it arrives with a velocity v_1. Conservation of energy requires

$$E_{1_0} = E_{1_a} + m_1V_a + \tfrac{1}{2}m_1v_1^2 = E_{1_a} \qquad (6.6)$$

with

$$V_a + \tfrac{1}{2}v_1^2 = 0 \qquad (6.7)$$

since a mass m_1 in the gravity potential V_a obtains a negative potential energy m_1V_a that compensates exactly for the kinetic energy.

Similarly,

$$E_{2_0} = E_{2_a} \quad \text{and} \quad v_1 = v_2 \qquad (6.8)$$

From Eqs. (6.7) and (6.8) we see that the energy levels are not changed by the free falling motion. *But* this does not tell us the complete story about the frequency of emitted radiation: We only know the absolute value of the velocity v_1, but the *direction of this velocity is unknown*. It depends on the detailed field distribution and trajectory, and the direction of the velocity is important for the Doppler effect.

It is absolutely necessary to *stop the falling clock and bring it to rest* in our original frame of reference; the clock must recover a zero velocity in this frame, and this means it must be stopped and strongly fastened to the heavy frame of reference (the laboratory at A) so as to send synchronous signals of fixed frequency all around (no Doppler effect).

In order to bring the clock to rest without perturbation, we must use *another type of force* (e.g. elastic force) to compensate the gravity forces. We actually put the clock on a table whose elastic stresses and strains keep the clock at rest despite its weight.

In Fig. 6.1 we assumed that the compensating forces depended on a potential V_s during the whole process of moving the clock, and that the total potential was

$$V_t = V + V_s = \epsilon \qquad (6.9)$$

which is very small, giving a transportation velocity v which is also extremely small. At point A the total potential is reduced exactly to zero, and the clock is brought to rest. The dashed curve in Fig. 6.1 represents V_t. At point A this total potential rises sharply, the clock is not supposed to be able to penetrate into the sphere.

This slow motion does not alter the energy levels E_1 and E_2 and the frequency ν remains unchanged for a local observer at rest. This is what happens in the Mössbauer effect, where elastic forces in a crystal lattice keep the atom at rest and compensate the gravity forces. The elastic forces of Mössbauer will absorb the recoil $h\nu/c$ due to emission of radiation.

The atom at rest at point A emits a photon $h\nu_0$ identical with the one emitted at rest at infinity [Eq. (6.5)]. This photon is observed at infinite distance at point B and one must notice that the photon is not sensitive to elastic forces, while its mass μ in motion makes it sensitive to gravity

$$\mu c^2 = h\nu \qquad (6.10)$$

While climbing the gravity field from A to B it loses energy, mass, and frequency. For a displacement $dr > 0$, let the potential increase be $dV > 0$

$$d(h\nu) = -\mu\, dV = -\frac{h\nu}{c^2}\, dV$$

or

$$\frac{d\nu}{\nu} = -\frac{dV}{c^2} \tag{6.11}$$

assuming c constant despite changing gravity.

We may assume that, in most practical cases, the potential increase from A to B is a small quantity and write

$$\frac{d\nu}{\nu} = \frac{V_a}{c^2}, \quad V_a < 0 \tag{6.12}$$

The photon's frequency is decreasing (red shift) and formulas (6.11) or (6.12) correspond to Einstein's prediction. We felt it necessary to discuss everything in detail, especially the role of elastic forces in the Mössbauer effect because most authors omitted one point or another, often reaching the correct result through incomplete reasonings.

The remarkable point is that the *local frequency, observed near the atomic clock at rest, is constant and does not depend on the local gravitational potential.* All our clocks, locally observed, remain strictly synchronous and unsensitive to the local potential V, but the frequency observed from a distance depends on the potential V and not on other sorts of potential energies.

Let us emphasize that nonstatic gravity fields may not derive from a potential V, thus leaving the question open.

The discussion may apply to photons or gravitons $h\nu$, all being uncharged and reacting only to gravity changes. Let us note that Einstein's original discussions, using a free falling clock and paying no attention to the problem of bringing it to rest, are incomplete.

3. Interpretations of the Gravity Red Shift

Experiments of Pound and Snider (1965) using the Mössbauer effect were usually considered as a verification of Einstein's prediction. The explanation given in the preceding section is quite different from Einstein's theory. This point being of importance, let us state the differences clearly:

a. We used a Euclidian space and quantized clocks corresponding exactly to the experimental device of Pound. We proved that such clocks gave locally a time definition *independent of gravity potential*; a more elaborate discussion in Section 6.4 will show that only a very small influence of the *gravity field* might be expected. It was essential in the discussion to assume a single frame of reference, at rest with respect to the heavy body creating a constant gravity field, and to specify that we should always use *clocks at rest* in this constant field. The heavy body, supporting our laboratory, defined our *preferred inertial* set of coordinates. And in this *frame of reference all clocks were exactly synchronous* locally whatever the gravity potential might be.

With this model, a change of *frequency occurs during the propagation* of the photon through the gravity field. We may be surprised by this statement and ask: How can this happen? Let us candidly admit that we do not know how to explain it. We have no model for such an effect. It rests directly upon Bohr's formula (6.3), which has never been explained in any " reasonable " way. We have to take it as an empirical result beyond our comprehension, but supported by an enormous amount of experimental observations.

b. Einstein used ideal clocks of an undefined structure, and attempted to discuss their behavior in a field of gravitation. This should not be considered as a criticism of Einstein: There was no knowledge at the turn of the century of how to build an ideal clock or exactly how it might behave.

The result, however, is that it is rather difficult to understand Einstein's discussion on the subject (Einstein, 1911; Einstein, 1924, pp. 100–107) and the best thing is to refer the reader to this fundamental paper. In a previous paper (Einstein, 1905; Einstein, 1924, p. 56) we find an excellent and complete discussion of the Doppler effect in restricted relativity, showing how the radiation frequency depends on the angle φ between the velocity v of the source and the direction of observation.

In the 1911 paper (Einstein, 1924, pp. 102–104) Einstein considers the case of a constant vertical field of gravitation, and assumes that this problem should be *equivalent* to another one with constant vertical acceleration. The conditions for such a principle of equivalence were never stated exactly, and have been very strongly criticized by many authors (e.g., Fock, 1964). In 1911

Einstein considered a clock starting with an initial velocity zero, falling vertically with a constant acceleration γ; after falling a height H, the clock reaches a velocity v. This clock emits radiations ν which, when observed from a fixed position, appear (by the Doppler effect) with a frequency

$$\nu' = \nu\left(1 + \frac{v}{c}\right) = \nu\left(1 + \frac{\gamma H}{c^2}\right)$$

Here Einstein uses the Doppler formula for $\varphi = 0$, thus assuming that the motion remains vertical and that radiation is observed from the original position of the falling clock. He does not care about any oblique Doppler effect or frequencies observed from angles $\varphi \neq 0$.

The whole discussion is really obscure. Read the following paragraph, for instance (Einstein, 1924, pp. 106, 107) and try to understand which clocks are of identical construction and which are not:

> For measuring time at a place which, relatively to the origin of coordinates, has the gravitational potential Φ, we must employ a clock which—when removed to the origin of co-ordinates—goes $(1 + \Phi/c^2)$ times more slowly than the clock used for measuring time at the origin of coordinates.

This is very strange: When building a clock in a laboratory it would not be enough for us to measure locally the gravity field in the laboratory, but we should know all the field distribution in the whole universe, up to infinite distances where $\Phi = 0$ in order to compute the local Φ in the laboratory!

The connection between the variable t in Einstein's equations and the time measured in our laboratories is far from clear. One may wonder why this very special problem of vertical motion was singled out for discussion. We may, for comparison, select another example: the *motion of an atomic clock along a Kepler ellipse around the sun*. The mass of the satellite may be large compared with the masses of photons emitted or absorbed. Furthermore, the trajectory remains unchanged when the mass of the satellite is modified, since it is moving in a " field of acceleration." At aphelion, the satellite is far away from the sun, the potential is small and the small velocity v_a is perpendicular to the radius r_a (Fig. 6.2). At perihelion, potential and velocity v_p are large. There is conservation of energy along the motion, so our energy levels

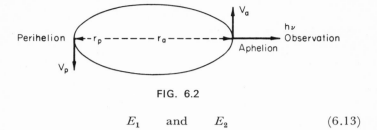

FIG. 6.2

$$E_1 \quad \text{and} \quad E_2 \tag{6.13}$$

remain unchanged. By selecting a very elongated ellipse, we may remove the aphelion as far as we want, and go to the limit of a parabola; hence we are sure that the energy levels are the same as at infinity

$$E_{1_0} \quad \text{and} \quad E_{2_0}$$

In order to eliminate the Doppler effect, we may choose to observe quanta $h\nu$ emitted along the radius, hence perpendicularly to the velocity, and we can state safely that the emitted photon is constant:

$$h\nu = h\nu_0 \tag{6.14}$$

independent of the gravity potential.

This is in complete agreement with our previous discussion. Instead of using clocks *at rest* we observe here clocks *keeping a constant distance from the sun* for a short time interval. Both procedures lead to identical results.

Einstein found it necessary to introduce a curved space–time and based his general relativity on this ideal notion. We did not feel the need for such curvature of the four-dimensional universe, because quantum conditions gave us a different answer. This situation was made even more obscure by some theoreticians who used both curvature of the universe and quantum theory, a mixture leading to hopeless confusion.

Let us try to summarize the situation. We use an *atomic clock*, whose properties are defined by the laws of quantum mechanics. As a result, we must assume *our clock to be at rest in an inertial frame of reference*, whether there exists a gravity field or not. This clock may (as we shall see in Section 6.4) be influenced by the gravity field, but it is unsensitive to gravity potential. All the clocks at rest in our inertial frame will give the same frequency definition, with or without gravity potential. The gravity red shift is only due to the motion of photons.

4. The Possibility of a " Gravi-Spectral " Effect

The discussion of Sections 6.2 and 6.3 predicts a frequency change due to the gravitational potential in a static problem.

We may also be surprised not to find anything similar to the Zeeman or Stark effects, where frequency of radiation depends on the *magnitude of the field, not the potential*. This simply means that such effects have been *ignored and overlooked*, but they should exist. We cannot discuss the problem for γ-rays, since the mechanism of these rays rests within the nucleus and is not known exactly. We shall select another problem, and consider an *atomic clock* using an optical frequency ν_0. The red shift due to the gravity potential is still given by formula (6.12), and we may use the theory of optical spectra for Stark effects. We assume that the forces acting upon the atom and maintaining it at rest in a crystal lattice act upon the nucleus of this atom, but not upon the surrounding electrons. This is obviously an oversimplified model used by Brillouin (1967) for the purpose of proving how some frequency changes due to the gravity field might occur.

The electrons surrounding the fixed nucleus are still feeling the local gravity field f_a [Eq. (6.4)] which gives equal and parallel forces on all individual electrons, just as a constant local electric field f_e would do:

$$ef_e = -m' \frac{\partial V_a}{\partial a} = f_a \tag{6.15}$$

where e and m' are the electronic charge and mass, and these forces must result in a very weak Stark-like multiplet. The order of magnitude of Stark multiplet splitting is

$$\Delta \nu_e = \frac{3ef_e n h}{8\pi^2 m' \mathcal{Z} e^2} = 137 \frac{3nef_a}{4\pi m' \mathcal{Z} c} \tag{6.16}$$

where $\mathcal{Z}e$ is the charge of the nucleus, n is an integer and $hc/2\pi e^2 = 137$. Let us now use Eq. (6.15) and consider the situation at point A of Fig. 6.1, where the atom is at a distance a from the center 0 of attraction M,

$$ef_e = f_a = -\frac{m'GM}{r^2} \tag{6.17}$$

where G is Newton's constant, hence

$$\Delta\nu_g = -137\frac{3n}{4\pi Zc}\frac{GM}{r^2} \tag{6.18}$$

for the " gravi-spectral " Stark-like effect.

Let us compare this new splitting due to gravity forces with the red shift produced by gravity potential (Eq. 6.12),

$$V_g = -\frac{GM}{r}, \qquad \frac{\delta\nu_{rel}}{\nu} = \frac{V_g}{c^2} = -\frac{GM}{rc^2} \tag{6.19}$$

The orders of magnitude can readily be compared

$$\frac{\Delta\nu_g}{\delta\nu_{rel}} = 137\frac{3n\lambda}{4\pi Zr}, \qquad \lambda = \frac{c}{\nu} = \text{wavelength} \tag{6.20}$$

The gravi-spectral effect $\Delta\nu_g$ is of the order of 137 (λ/r) times the relativistic red shift $\delta\nu_{rel}$. It could be observed only for very short distances r from the center of attraction M.

This short discussion proves that direct effects of the *gravity* field should exist, but may be practically very difficult to observe. Einstein's theory and quantized atomic theories both ignored this possibility, and experimental checks would be important.

REFERENCES

Brillouin, L. (1967). *Proc. Natl. Acad. Sci. U.S.* **67**, 1529.
de Donder, T. (1921). " La gravifique Einsteinienne." Paris.
Einstein, A. (1905), *Ann. Physik* **17**, 891.
Einstein, A. (1911). *Ann. Physik* **35**, 898.
Einstein, A. (1924). " Principle of Relativity " (W. Pennett and G. B. Jeffery, trans.). Dover, New York.
Fock, V. (1964). " The Theory of Space, Time and Gravitation." Pergamon Press, Oxford.
Lanczos, C. (1922). *Physik Z.* **23**, 537.
Pound, R. V. and Snider, R. L. (1965). *Phys. Rev.* **140**, 788.

Chapter 7

A Gravistatic Problem with Spherical Symmetry

1. A New Approach to an Old Problem

We discussed in Section 4.7 the famous Schwartzschild problem according to Einstein's theory. We should explain that this problem is usually stated in a curious way: One considers the case of a " point mass at rest " and looks for a solution exhibiting spherical symmetry in space with no time dependence. This does not tell anything about *boundary conditions*. Later in the discussion it is assumed that at large distances the metric tensor should correspond to *Euclidian vacuum*, thus specifying boundary conditions at infinity; but no specific conditions are given for small distances from the origin. A complete statement of the problem should include the conditions on a small sphere *a* enclosing the origin (which is a singular point). In Schwarzschild's discussion, the r_0 value is not given, and the mass of the " point mass " is only introduced at the end of the computation in order to obtain a potential decreasing in Gm/r at large distances r, where G is Newton's constant. This is one of the reasons why there remains so much uncertainty in Schwarzschild's solution. In our discussion we shall see that Schwarzschild's mass m represents the mass of the center point plus mass densities in the field.

What we intend to do now is to start from the classical problem of gravitation around a sphere of given mass and given radius; then we shall examine how this solution must be modified to include the role of mass-density distribution in the space around the sphere resulting from the energy density in the field and the mass-energy relation. This should give us a reasonable generalization of the classical solution, in which the physical meaning of all quantities can be correctly understood. Einstein's equations are not used.

2. Gravistatics Compared to Electrostatics

In the static problem with spherical symmetry, we immediately notice that we can put aside the variable t, which plays no role. Our discussion of Chapter 6 enables us to use *a preferred frame of reference*, where the origin of coordinates is at rest at the center of the spherical mass; in this frame of reference, all the atomic clocks maintain a perfect simultaneity, and we have a *single* time definition throughout. This does not interfere with the well-known gravity red shift (discussed in Chapter 6) that results from the motions of photons of mass $h\nu/c^2$ through the gravity field.

We may now compare similar problems in gravitation and in electrostatics. This was done by Brillouin and Lucas (1966) and Mannheimer (1966). In both cases we start with forces decreasing in r^{-2} with the distance r. We have the following laws:

Coulomb's law for charges Q_1 and Q_2, dielectric power e is given by:

$$\mathbf{f} = \frac{Q_1 Q_2}{er^2}\,\mathbf{r}^0 \tag{7.1}$$

Newton's law for masses M_1 and M_2, Newton's constant G is written:

$$\mathbf{f} = -G\,\frac{M_1 M_2}{r^2}\,\mathbf{r}^0, \qquad G = 6.66 \times 10^{-8}\ \text{CGS} \tag{7.2}$$

The notation $\mathbf{r}^0$ represents a unit vector in the direction $\mathbf{r}$. Both formulas (7.1) and (7.2) are identical if we assume

$$e = -1/G = -1.5 \times 10^7 \tag{7.3}$$

Newton's attraction corresponds to a large negative dielectric constant. We repeatedly stated the need for using positive or negative masses, since we may have positive or negative energies and we must keep the mass–energy relation unchanged:

$$E = Mc^2 \tag{7.4}$$

There is only one coefficient M for each particle; this coefficient is playing its role in the relation above and also in the inertial law

$$\mathbf{f} = M\gamma \tag{7.5}$$

where γ is the acceleration. Figure 7.1 will show the signs of $\mathbf{f}$ and acceleration γ in a variety of conditions for two masses M_1 and M_2 interacting together.

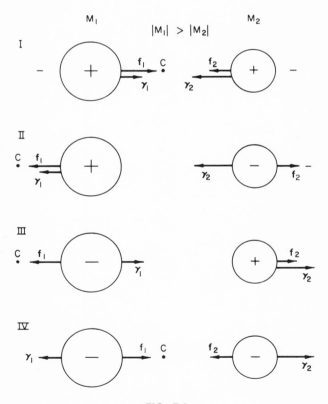

FIG. 7.1

Figure 7.1 shows how much a " mass plasma " would differ from an electric plasma; attractions and repulsions would not lead to the same types of mixtures in both cases. Note that the acceleration is the same for positive or negative moving objects, in accordance with the classical idea of a " field of acceleration."

3. Some Essential Formulas from Electrostatics

Let us now rewrite a few classical formulas which we may use for electrostatics or gravistatics:

$$\mathbf{F} = -\nabla V \tag{7.6}$$

$$\mathbf{D} = e\mathbf{F} \tag{7.7}$$

$$\nabla \cdot \mathbf{D} = 4\pi\rho_0 \tag{7.8}$$

where V is the static potential, $\mathbf{F}$ is the field, $\mathbf{D}$ is the displacement and ρ_0 is the mass or charge density. The energy density in the field is given by

$$\mathscr{E} = \frac{1}{8\pi}(\mathbf{F} \cdot \mathbf{D}) \tag{7.9}$$

or, provided e is strictly constant

$$\mathscr{E} = \frac{eF^2}{8\pi} = \frac{D^2}{8\pi e} \tag{7.10}$$

Let us now consider a point charge Q (or a point mass M)

$$\mathbf{D} = \frac{Q}{r^2}\mathbf{r}^0, \qquad \mathbf{F} = \frac{Q}{er^2}\mathbf{r}^0, \qquad V = \frac{Q}{er} \tag{7.11}$$

The formula for electrostatic energy density (7.9) was already used in Chapter 2, where it was shown that the volume integral of this density did yield the classical potential energy. The difference between electrostatics and gravistatics is that a point charge Q may actually exist, while a point mass M is practically impossible. Every mass M is surrounded by an atmosphere of *mass densities* resulting from the energy densities in the field [Eqs. (7.4) and (7.9)].

Let us first show how formulas (7.11) must be completed and corrected when the mass can no longer be considered as infinitely small. We do not want to discuss what may happen within the sphere a; this inside problem is a different story and should be put aside. So, we choose to consider an *empty spherical shell* or bubble of mass M_0. There is no field inside the bubble if the *mass M_0 is uniformly distributed* on the sphere, hence no perturbation to the usual theory within the bubble. Outside the bubble, Eqs. (7.11) can be used as a first approximation and yield an energy density and a mass density $\mathcal{U}_g$ [Eqs. (7.9) and (7.10)].

$$\mathcal{E} = \frac{1}{8\pi} (\mathbf{F} \cdot \mathbf{D}) = -G \frac{M_0{}^2}{8\pi r^4} = \mathcal{U}_g c^2, \qquad \text{for } r \geqslant a \quad (7.12)$$

by the mass-energy relation (7.4). We thus discover an atmosphere of negative mass all around the bubble M_0 of radius a. This atmosphere surrounding M_0 is *always negative*, whatever the sign of M_0 might be. The total mass M_g distributed in the field is directly obtained by integration for the whole space

$$\mathbf{M_g} = -\frac{G}{c^2} \frac{M_0{}^2}{2a} \qquad (7.13)$$

This formula corresponds to the one giving the electromagnetic mass of an electron and represents a *very small* relative correction when $| GM_0/2c^2a |$ is small. The mass M_0 could be measured only by instruments located very close to the bubble. At large distances r, we measure a total mass

$$M_t = M_0 + M_g + \cdots = M_0 \left(1 - \frac{GM_0}{2c^2a} + \cdots \right), \qquad \left| \frac{GM_0}{2c^2a} \right| \ll 1$$
$$(7.14)$$

If $| GM_0/2c^2a |$ happens to be large, we have to consider higher approximations. Let us immediately notice the *nonlinear character* of gravistatics and the *dissymmetry* between *positive and negative masses*.

4. Complete Gravistatic Field with Surrounding Mass-Density Distribution

We may easily discover the fundamental laws of gravistatics. Let us start from the energy-density and mass-density formulas (7.3), (7.4), (7.10):

$$\mathscr{U}_g = -\frac{GD^2}{8\pi c^2} \tag{7.15}$$

and combine it with (7.8) to obtain

$$\mathbf{V} \cdot \mathbf{D} = 4\pi \mathscr{U}_g = -\tfrac{1}{2}gD^2 \quad \text{with} \quad g = \frac{G}{c^2} \tag{7.16}$$

This is our *fundamental nonlinear law for gravistatics.* Let us use our condition for spherical symmetry, assuming $\mathbf{D}$ to be D_r along the radius:

$$\left(\frac{1}{r^2}\right)\frac{d}{dr}\,(r^2 D_r) = -\tfrac{1}{2}gD_r{}^2 \tag{7.17}$$

We note that $r^2 D_r$ represents the total mass M_r within a sphere r [see Eq. (7.11)]

$$\frac{dM_r}{dr} = -\tfrac{1}{2}g\,\frac{M_r{}^2}{r^2}, \qquad M_r = r^2 D_r \tag{7.18}$$

Let us use the reduced mass m_r, which was defined in Eq. (4.3)

$$m_r = \frac{G}{c^2}\,M_r = gM_r = gr^2 D_r \tag{7.19}$$

and we have the equation

$$\frac{dm_r}{dr} = -\frac{m_r{}^2}{2r^2} \tag{7.20}$$

Integration yields

$$\frac{1}{m_r} = -\frac{1}{2r} + \frac{1}{2\alpha} \tag{7.21}$$

where α is an integration constant, hence

$$m_r = \frac{2r\alpha}{r - \alpha}, \qquad D_r = \frac{2\alpha}{r(r - \alpha)}$$

At large distance, we obtain Newton's field for the total mass m_t (bubble mass m_0 plus field mass m_f)

$$m_t = m_0 + m_f = 2\alpha, \qquad r \gg \alpha$$

but (7.21) yields

$$m_0 = \frac{2a\alpha}{a - \alpha} = \frac{m_t}{1 - m_t/2a} \tag{7.22}$$

This is the correct answer, while our Eq. (7.14) gave only a *first approximation*:

$$m_0 = \frac{m_t}{1 - m_0/2a} \tag{7.14}'$$

According to the correct formula (7.22) we note that the theory diverges when

$$a = m_t/2 \tag{7.23}$$

We shall discuss this result, which corresponds to conditions for *gravitational collapse*, in the next section.

5. Discussion

Before starting the discussion we must make an important remark: The *mass distributed in the field is always negative*, since gravitation corresponds to a negative dielectric constant [Eq. (7.3)]. Hence:

$$m_t - m_0 = m_f < 0 \tag{7.24}$$

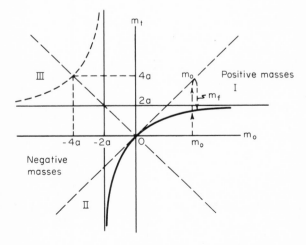

FIG. 7.2

where m_f is the mass in the field. Using Eq. (7.22) we obtain

$$m_t - m_0 = -\frac{m_0 m_t}{2a}, \quad \text{hence} \quad m_0 m_t > 0 \qquad (7.25)$$

The mass m_0 of the bubble core and the total mass m_t always have similar signs. Fig. 7.2 visualizes the relation between m_0 and m_t; the curve is an equilateral hyperbola, but only the following branches have a physical meaning:

branch I $m_0 > 0$, $m_t > 0$,   positive masses

branch II $m_0 < 0$, $m_t < 0$,   negative masses

while branch III, with $m_0 < 0$ and $m_t > 0$, has no physical meaning.

The strong dissymmetry between positive and negative masses is very striking. For *positive masses* we see that the total reduced mass m_t cannot exceed $2a$

$$m_t \leqslant 2a, \quad \text{for} \quad m_0 \to \infty \qquad (7.26a)$$

This is again condition (7.23).

For *negative masses* we have a very different situation:

$$m_t \to -\infty, \quad m_0 \geqslant -2a \qquad (7.26b)$$

These curious limitations require closer examination.

Let us now come back to condition (7.23) giving the critical relation between mass m_t and the radius a of the bubble. It indicates gravitational collapse when the *radius a* equals *one-half of the total mass m_t* (the mass of the central bubble plus the distributed mass in the surrounding field).

We may compare this result with the values computed from Einstein's theory, which were summarized in Chapter 4 [Eqs. (4.3–8)]. Einstein does not discuss the distribution of mass between the central core and the field, and his mass m corresponds to our m_t.

Furthermore, Einstein does not specify the type of coordinates to be chosen, and even takes pride at this lack of definition. We emphasized the need for such a definition before making any attempt at an experimental check, and we assumed isotropic Euclidean space. This corresponds to formulas (4.5) and (4.6), where we also obtain a critical radius of $\frac{1}{2}m$. There is complete agreement between our elementary discussion and Einstein's solution for Euclidian space.

The practical discussion of this chapter answers the question raised in Chapter 4 and strongly suggests that an isotropic Euclidean space with a variable light velocity should be the model closest to experimental physical conditions.

Fock's preferred frame of reference (4.7) and (4.8) does not agree with our physical discussion.

REFERENCES

Brillouin, L. and Lucas, R. (1966). *J. Phys. Radium* **27**, 229.
Mannheimer, M. (1966). *Ann. Phys. (Paris)* **1**, 189.

Chapter 8   **Remarks and Suggestions**

1. The Meaning of a Spectral Line

The question as to the meaning of a spectral line has been asked often and different answers have been given. A spectral line defines a wavelength in optics, and for many years the only methods of observation were based on interference phenomena. Optical textbooks spoke of " frequency " and gave figures in reciprocal centimeters because of the uncertainty of the light velocity c. As we saw in Chapter 3, spectral lines are also now being used to define an actual frequency in reciprocal seconds, and we emphasized in the Introduction the very ambiguous present situation. The unit of length is officially based on a spectral line of krypton-86, while the second of time is defined by a spectral line of cesium. So, if one wants to measure the velocity of light c one just has to observe the ratio of frequencies (or wavelengths) of krypton and cesium! This is a curious, illogical statement in science.

We discussed the gravity red shift in Chapter 6, assuming that h and c were actual constants. But many authors agree with Einstein and claim that the velocity of light c depends upon the gravity potential in every static problem where such a potential can be defined; hence we must inquire: Is the red shift due to an increased wavelength (at constant frequency) or shall it be construed as a decreased frequency?

We stated our results in Chapter 6 as if they corresponded to an actual change of frequency, and we found this point of view very difficult to explain physically. It would be much easier to understand a change in the light velocity c, resulting in a changed λ at a constant frequency v.

This, independently of any theory, is actually an experimental fact and was recently verified by Shapiro (1968) in a brilliant series of observations where a very accurate laser beam was reflected on the planet Mercury and came very close to the sun on its way back to earth. The beam traveling near the sun was propagated more slowly, and excess delays of 125 microseconds were clearly observed. This experiment clearly shows that the velocity of light in the neighborhood of the sun is smaller than at large distances.

In connection with these problems, we must recall a very interesting paper by Lucas (1966), where he discusses how a number of physical properties might be modified by a change in the gravity potential. Lucas makes the very interesting assumption that h/c^2 is an *invariant*, thus keeping constant the ratio of mass to frequency. Such an assumption has the significant advantage of keeping unchanged our discussion of Section 6.2 about Pound's experiments.

2. General Gravitation Theory and Experiments

After developing his general relativity theory, Einstein predicted some effects that might be tested by observations. Many attempts were made since that time, and few practical results were obtained. First of all, let us state clearly that such predictions are not specifically tied up to Einstein's theory; very similar predictions with only slight differences of order of magnitudes obtain similar results for *any computation including the mass–energy relation*.

For instance, Einstein predicts the *deflection of a light ray* passing near the surface of the sun, but we obtain a similar result if we consider a light ray as a beam of photons hv with masses hv/c^2. Only the numerical coefficient is different, and Einstein's prediction is twice as large as that in the computation with photons. Here the experimental results are actually very poor, with errors of 100% magnitude; a detailed discussion of older results may be found in Chazy's book (1930), and more recent experiments were no better; looking candidly at these observations, one feels that very large sources of error are obviously playing a substantial role, and our present knowledge of the turbulent flow in the solar atmosphere yields the most probable explanation. The Shapiro (1968) experiment is certainly safer than the deflection of light rays.

Let us emphasize the importance of the *solar wind* that corresponds to ten million tons of matter annihilated per second and radiated away!

The advance of the perihelion of Mercury (43 seconds per century) was hailed as a wonderful check with a theoretical prediction of $42''$ 6, but here again let us refer to Chazy (1930) who found a number of other examples in the solar system where Einstein's predictions conflict with experiments. It is hard to believe seriously in a coincidence of less than one second for Mercury, while so many other examples give large errors and even opposite signs! Let us here candidly admit that there must be many other unknown factors involved. The computations of Chazy refer to the motions of perihelions of four planets and similar motions for a number of satellites orbiting around planets (e.g., the moon). Errors of at least five seconds per century seem to be the inevitable limit in these very difficult computations. Einstein's theory yields about $\frac{1}{6}$ of the advance of perihelion of Mars and practically nothing for Venus. Let us add that Dicke's discovery of the oblate shape of the sun leads to perturbations that definitely destroy the agreement about Mercury. The question cannot be considered completely settled.

3. Bridgman's Reappraisal of Relativity

All the discussions of this book were strongly influenced by Bridgman's ideas and his emphasis of the constant need for interrelation between theory and experiment. This viewpoint coincided with that of the present author, and was in complete agreement with the traditional thinking of a large school of French scientific philosophers, especially Berthelot (1863), Curie (1908), and M. Brillouin (1935).

We must now reread and quote from the last book of Bridgman (1962), published after the scientist's untimely death, which contains a very thorough reappraisal of relativity from a " sophisticated primary " viewpoint. On many occasions, we discover that Bridgman came very close to our line of discussion, and we hope to show that this book may be considered an extension of Bridgman's methods and ideas.

Bridgman starts with the Lorentz equations in his first chapter, for reasons of " convenience," and immediately states his belief that the Lorentz formulas represent only a " practical characterization " of special relativity. We agree with this distinction and tried to distinguish sharply between both points of view (the classical and ours) when we specified the special postulate required to get rid of first-order Doppler effects and keep only the second-order effects of Lorentz (see Chapter 5).

Bridgman very clearly discusses the problem of setting clocks at a distance and " spreading time over space." Regarding a penetrating appraisal of Reichenbach (1958), it is interesting to note how Bridgman maintains the validity of the naïve and old method for setting clocks at a distance by transporting a clock from one place to another. He shows how it can be correctly defined to agree completely with the discussions of Einstein and Reichenbach.

Let us also note how cautiously Bridgman speaks of Galilean frames (pp. 78–79):

> A Galilean frame is a rigid physical scaffolding, to which a coordinate system can be attached.... The members of the frame are free from internal stresses.... Associated with the particle (at the origin), *there has to be some mass* which serves as the origin of the primitive set.... If we expect to use the framework as an anchorage for the arbitrary forces we want to apply to the various particles to induce in them any desired state of motion, then we shall *obviously have to make the frame massive as well as rigid.* " Massive " means much heavier than any of the particles we expect to put into interaction with [the earth, with some small corrections].

Bridgman's statement completely agrees with the viewpoint we presented in Chapter 4 of this book.

There are many illuminating remarks in Bridgman's booklet, and every physicist will enjoy reading it and commenting on its suggestions. Among many other bright ideas, let us point to a very curious one (pp. 159, 160). Bridgman compares the electromagnetic theory, with its *constant light velocity c*, and classical mechanics, where a " free " moving massive particle, in a Galilean frame of reference, maintains a *constant given velocity v* for any time. Both results strike him as analogous and simply wonderful!

He wonders whether there might be some deep similarity, *something analogous to the electromagnetic field equations, but applicable to inertial matter!* He is tempted to believe that some new physical effect may have escaped detection. In electromagnetism, one needs two vectors **E** and **H**; assuming the inertial **E** to correspond to gravitation, what should be the role of the inertial **H**?

Bridgman does not elaborate, but despite his proverbial cautiousness, he is not afraid of stating such a fantastic suggestion, and this situation is worth scrutinizing. This may serve to introduce Carstoiu's investigation in the following section.

4. Carstoiu's Suggestions for Gravity Waves

Carstoiu (1969) starts from the discussion of Brillouin and Lucas (1966) that was restated and corrected in Chapter 7. We emphasized the startling similarity between electrostatics and equations of a static gravity field **F** (gravistatics). In order to discuss non-static problems, Carstoiu assumes the existence of a *second gravitational field* called the *gravitational vortex* Ω; both fields are supposed to be coupled by equations similar to Maxwell's equations, and obtain a propagation velocity c equal to the velocity of light.

As is well known, Maxwell's equations contain two constants, the dielectric constant ε and the permeability μ, related by the condition

$$\varepsilon\mu c^2 = 1 \qquad (8.1)$$

thus yielding a velocity c for wave propagation.

Accordingly, Carstoiu introduces two gravitation constants ε_g and μ_g. Let us take for the ε_g the value we selected in Eq. (7.1):

$$\varepsilon_g = -1/G \qquad (8.2)$$

where G is Newton's gravitation constant. This leads to selecting:

$$\mu_g = -G/c^2 \qquad (8.3)$$

in order to satisfy condition (8.1). Actually, Carstoiu uses a different set of unities that result in replacing our G by $4\pi\gamma$, and he calls **G** the gravity field that we call **F**. Rewriting Maxwell's equation, Carstoiu obtains:

$$\text{curl } \mathbf{F} = -\frac{\partial \Omega}{\partial t}, \qquad \text{curl } \Omega = \frac{1}{c^2}\frac{\partial \mathbf{F}}{\partial t} - \frac{G}{c^2}\mathbf{J_g} \qquad (8.4)$$

$$\mathbf{V} \cdot \mathbf{F} = -G\rho_g, \qquad \mathbf{V} \cdot \Omega = 0$$

where ρ_g is the mass density, $\mathbf{J_g}$ the gravitational current, and Ω the gravitational vortex. Carstoiu discusses the possible role of his gravitational vortex on the stability of rotating masses and a variety of problems in cosmogony.

Let us consider again the nonlinear problem discussed in Chapter 7; we find a similar situation in the propagation equation. The energy density in the field contains terms in $|\mathbf{F}|^2$ similar to Eq. (7.10) and also terms in $|\Omega|^2$ for gravitational vortex. Energy density yields new mass density, hence an additional ρ_{add} term:

$$\rho_{add} = -\frac{1}{Gc^2}\frac{|\mathbf{F}|^2 + c^2|\Omega|^2}{8\pi} \qquad (8.5)$$

and we end up with nonlinear equations for gravity propagation. This ρ_{add} always represents negative mass, as noted in Chapter 7. This extension of Carstoiu's theory opens a large field for investigation. What is the meaning of the gravitational vortex and what sort of role could it play? How and where could it be observed? Let us only state, for the moment, that this new line of investigation may not be very far from Einstein's equations of gravity propagation, since Einstein's equations have been reduced by some authors to a schema similar to (8.4). The reader can refer to Carstoiu's papers for further investigation.

Let us indicate here that the similarity of Carstoiu's gravitational equations with Maxwell's electromagnetism leads to some curious suggestions: Both types of waves are transverse and propagate with the same velocity c, a coincidence that should facilitate a strong interaction if there happens to be any possibility for intercoupling; and such a possibility is immediately obvious. Electromagnetic fields create an energy density, according to a classical formula

$$\mathscr{E}_{EM} = \frac{\varepsilon \mathbf{E}^2}{8\pi} + \frac{\mu \mathbf{H}^2}{8\pi} = \rho_{EM,\,add}\, c^2. \qquad (8.6)$$

where $\mathbf{E}$ and $\mathbf{H}$ represent the electric and magnetic fields, respectively. This electromagnetic *energy density* $\mathscr{E}_{EM}$ represents a positive mass-density $\rho_{EM,\,add}$ to be added to our previous negative ρ_{add} of Eq. (8.5), and this mass-density distribution in any type of electromagnetic field must generate new gravitational fields. Thus, we have a very clear indication of a simple coupling between electromagnetism and gravitation, a problem open for further discussion.

After writing the papers just discussed here, Carstoiu discovered a very extraordinary note of Heaviside (1893, 1950), where he suggests for gravitation a set of equations very similar to Maxwell's electromagnetic equations and Carstoiu's formulas. Heaviside shows that these equations require the introduction of a second field, analogous to the magnetic force; this is Carstoiu's vortex Ω. It is very strange that such an important paper had been practically ignored for so many years, but the reader may remember that Heaviside was the forgotten genius of physics, abandoned by everybody except a few faithful friends.

5. Wanted: A Graser!

We are reaching the end of this essay, and after discussing so many problems, both theoretical and experimental, we must come to a conclusion: What Einstein's genius could not achieve, we doubt that any modern scientist, even another genius, can achieve. We have accumulated since the beginning of this century an enormous amount of knowledge; most of the discoveries were of an experimental nature, and theory could proceed only after a firm basis of empirical data had been built. When we consider gravity, its nature, its propagation, we must candidly admit that the progress was almost nil. We know little more than a century ago, " because measurable effects happen to be incredibly small "; since we cannot change their order of magnitude, the only thing we can do is to change our procedure of observation.

Radio was at a standstill until amplification was invented by De Forest; optics progressed slowly until Townes invented masers and lasers were built by Kastler. Who will now build a *graser*, a powerful amplifying device for gravity waves? When and if our observations reach a power one million times greater than today, we should be able to measure gravity waves, their frequencies, their velocities, and how they propagate.

We shall know whether these waves be longitudinal (like sound waves in gases) or transverse (Maxwell's equations), or mixed, tensorial waves. Many physicists had in mind longitudinal waves, while Heaviside, Bridgman, and Carstoiu suggested transverse waves similar to electromagnetic waves. We should be able to answer at least this question!

With a *graser* we might ascertain whether the velocity of gravity waves actually equals the velocity of light. If gravity waves happened to propagate more slowly than c, we should observe *gravity shock waves* for all particles moving with velocities close to c, and there are many such particles. A number of important problems could be solved, and many new roads open. Such a discovery would spark a big new chapter in physics, and engineers might even build gravity transmitters and receivers competing with radio! In scientific research, there is no substitute for observation. What we need is a *graser*!

REFERENCES

Berthelot, M. (1863) in E. Renan, " Dialogues philosophiques," p. 193. Calman-Levy, Paris.
This long letter of Berthelot to Renan is a remarkable sketch of the meaning and methods of natural philosophy. To understand it clearly, remember that " science idéale " means theory and " science positive " means experimental knowledge.

Bridgman, P. W. (1962). " A Sophisticate's Primer of Relativity." Wesleyan Univ. Press, Middletown, Connecticut.

Brillouin, M. (1935). " Jubilé scientifique," 2 vols. Gauthier-Villars, Paris.

Carstoiu, J. (1969). *Compt. Rend.* **268**, 201–263.

Chazy, J. (1930). " Théorie de relativité et mécanique céleste," 2 vols. Gauthier-Villars, Paris.

Curie, P. (1908). " Œuvres de Pierre Curie." Gauthier-Villars, Paris.
Read especially the articles on symmetry.

Heaviside, O. (1893, reprinted 1950). " Electromagnetic Theory." Dover, New York.
See especially Appendix B, pp. 115–118; also the quotation in the Introduction of this book.

Lucas, R. (1966). *Compt. Rend.* **262**, 853.

Reichenbach, H. (1958). " The Philosophy of Space and Time." Dover, New York.

Shapiro, I. I. (1968). *Sci. Am.* **219** (1), 28.

Index